Gasification

Gasification

Editor

Navneet Kumar

Gasification
Edited by **Navneet Kumar**

Printed in 2017

ISBN: 978-1-68117-388-7

Library of Congress Control Number: 2015941577

© 2016 by

SCITUS Academics LLC,
616, Corporate Way, Suite 2, 4766,
Valley Cottage, NY 10989

www.scitusacademics.com

Contents

Preface vii

Chapter 1   Effect of Gasifying Medium on the Coal Chemical Looping
             Gasification with CaSO4 as Oxygen Carrier 1
             Yongzhuo Liu, Weihua Jia, Qingjie Guo, and Hojung Ryu

Chapter 2   Cao-based Chemical Looping Gasification of Biomass for
             Hydrogen-enriched Gas Production with in Situ CO$_2$
             Capture and Tar Reduction 23
             Jakkapong Udomsirichakorn, Prabir Basu, P. Abdul Salam,
             and Bishnu Acharya

Chapter 3   Bubbling Fluidised Bed Gasification of Wheat Straw–Gasifier
             Performance Using Mullite as Bed Material 45
             Seán T. Mac an Bhaird, Phil Hemmingway, Eilín Walsh,
             Amado L. Maglinao, Sergio C. Capareda, and Kevin P. McDonnell

Chapter 4   Assessment of Chemical Looping-Based Conceptual Designs
             for High Efficient Hydrogen and Power
             Co-generation Applied to Gasification Processes 69
             Calin-Cristian Cormos, Ana-Maria Cormos, and Letitia Petrescu

Chapter 5   Model Design of a Class of Moving-Bed Tubular
             Gasification Reactors 103
             Ulises Badillo-Hernandez, Luis Alvarez-Icaza, and Jesus Alvarez,

Chapter 6   A New Method to Calculate Kinetic Parameters Independent
             of the Kinetic Model: Insights on CO$_2$ and Steam Gasification 151
             Arturo Gomez and Nader Mahinpey

Chapter 7   CFD–DEM Simulation of Biomass Gasification with Steam in a
             Fluidized Bed Reactor 189
             Xiaoke Ku, Tian Li, and Terese Løvås

Citations 227

Index 231

Preface

Gasification provides an excellent overview of current technologies for the gasification of coal, oil, gas, biomass and waste feed stocks. Starting from the basic theory, it reviews the potential feed stocks and their suitability for different types of gasification process. Commercial and near-commercial processes are described individually and various features discussed in detail. There is a comprehensive review of contaminants in synthesis gas as well as of gas treating processes. Gasification, the key technology enabling the production of biofuels from all viable sources-- some examples being sugar cane and switchgrass. This versatile resource not only explains the basic principles of energy conversion systems, but also provides valuable insight into the design of biomass gasifiers. The author provides many worked out design problems, step-by-step design procedures and real data on commercially operating systems. After fossil fuels, biomass is the most widely used fuel in the world. Biomass resources show a considerable potential in the long term if residues are properly handled and dedicated energy crops are grown.

Editor

Effect of Gasifying Medium on the Coal Chemical Looping Gasification with CaSO$_4$ as Oxygen Carrier

Yongzhuo Liu[1], Weihua Jia[1], Qingjie Guo[1], and Hojung Ryu[2]

[1]College of Chemical Engineering, Qingdao University of Science & Technology, Key Laboratory of Clean Chemical Processing Engineering of Shandong Province, Qingdao 266000, China

[2]Climate Change Technology Research Division, Korea Institute of Energy Research, Daejeon 305-343, Republic of Korea

ABSTRACT

The chemical looping gasification uses an oxygen carrier for solid fuel gasification by supplying insufficient lattice oxygen. The effect of gasifying medium on the coal chemical looping gasification with $CaSO_4$ as oxygen carrier is investigated in this paper. The thermodynamical analysis indicates that the addition of steam and CO_2 into the system can reduce the reaction temperature, at which the concentration of syngas reaches its maximum value. Experimental result in thermogravimetric analyzer and a fixed-bed reactor shows that the mixture sample goes through three stages, drying stage, pyrolysis stage and chemical looping gasification stage, with the temperature for three different gaseous media. The peak fitting and isoconversional methods are used to determine the reaction mechanism of the complex reactions in the chemical looping gasification process. It demonstrates that the gasifying medium (steam or CO_2) boosts the chemical looping process by reducing the activation energy in the overall reaction and gasification reactions of coal char. However, the mechanism using steam as the gasifying medium differs from that using CO_2. With steam as the gasifying medium, parallel reactions occur in the beginning stage, followed by a limiting stage shifting from a kinetic to a diffusion regime. It is opposite to the reaction mechanism with CO_2 as the gasifying medium.

GRAPHICAL ABSTRACT

- Graph 1. Chemical looping gasification using $CaSO_4$ as oxygen carrier is illustrated above. Coal gasification first occurs while oxygen carrier reacts with coal or syngas and is reduced to CaS. Then the as-reduced oxygen carrier is transferred to the air reactor and re-oxidized by air. The strong exothermic oxidation reaction can provide sufficient heat for coal gasification, which is an endothermic reaction.

- Graph 2. In a CO_2 atmosphere, the gasification of coal char and the reaction between coal char and $CaSO_4$ start at approximately the same temperature. However, the gasification of coal char in steam occurs at a lower temperature, with the reaction between coal char and $CaSO_4$ divided into two evident stages.

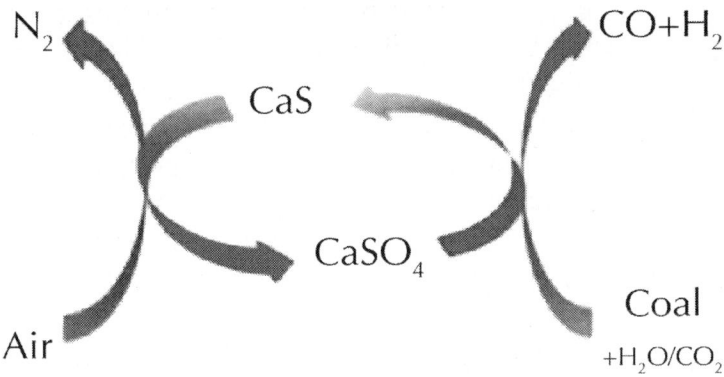

INTRODUCTION

Chemical looping technology, using lattice oxygen instead of molecular oxygen to avoid fuel directly contacting with the air, provides a new means of utilizing fossil fuels. Two approaches for direct chemical looping combustion (CLC) of solid fuel, characterized by inherent CO_2 separation and high efficiency, have been proposed [1] and [2]: *in-situ* gasification of solid fuel with H_2O or CO_2 as fluidization agent and chemical looping with oxygen uncoupled, where solid fuel is burned with gaseous oxygen released by the oxygen carrier in the fuel reactor [3]. However, unburned compounds (CO, H_2, CH_4, and other volatiles) from devolatilization or char gasification in direct CLC will appear in the combustion gases, which contain primarily CO_2 and H_2O. Chemical looping gasification (CLG) can overcome this problem, because the desired product is syngas, mainly containing CO, H_2, and CH_4. Moreover, the syngas is an indispensable product in the chemical, oil, and energy industries.

Syngas generation through the chemical looping process has been investigated by several researchers [4], [5] and [6]. Fan *et al.* [4] developed three chemical looping processes to convert carbonaceous fuels into products such as hydrogen, electricity, and synthetic fuels. Mattisson and Lyngfelt [7] proposed two chemical-looping reforming processes for syngas generation from hydrocarbons. He *et al.* [8] and de Diego*et al.* [9] and [10] investigated chemical looping reforming of methane to produce syngas using different metal oxide oxygen carriers. However, instead of producing syngas using lattice oxygen as a direct oxygen source during solid fuel gasification, they obtained syngas either from steam/CO_2 reforming using a reduced metal oxide (*e.g.*, Fe) or by chemical looping reforming of gaseous hydrocarbons.

Andrus *et al.* [11] proposed a chemical looping technique using a calcium-based, hybrid combustion–gasification approach for producing electricity, syngas or hydrogen from coal by controlling the air-to-coal ratio (*i.e.*, the oxygen carrier-to-coal ratio in the fuel reactor). If sufficient oxygen in the oxygen carrier reacts with the coal, carbon and hydrogen in the coal leave the fuel reactor as CO_2 and H_2O. If insufficient oxygen reacts with the coal, syngas CO/H_2 is generated; hydrogen will be produced if coupling with the water shift reaction and CaO carbonization reaction. However, further results of this process for direct syngas generation have not been reported in literature.

Chemical looping gasification of solid fuel for direct syngas generation shares the same basic principles as CLC. The process uses oxygen carriers to transfer oxygen and heat to the fuel reactor for coal gasification. However, CLG process produces the syngas with insufficient lattice oxygen provided to the solid fuel. As illustrated in Fig. 1, in the fuel reactor, coal is gasified; oxygen carrier reacts with coal or syngas and is reduced to its reduction state. Then, the as-reduced oxygen carrier is transferred to the air reactor and re-oxidized. The strong exothermic oxidation reaction can provide sufficient heat for coal gasification, which is an endothermic reaction. Compared with traditional coal gasification, costly air separation unit devices can be avoided in CLG by employing lattice oxygen instead of molecular oxygen. Investigation on the reaction mechanism of the system involving oxygen carrier and coal, with or without a gasifying medium, is vital to understand the CLG process. Compared with the traditional metal oxide oxygen carriers, oxygen carrier $CaSO_4$ possesses a high oxygen capacity, while its reduced state CaS has a high reaction enthalpy with the molecular oxygen in the air. $CaSO_4$ is regarded as a promising candidate of oxygen carrier [12].

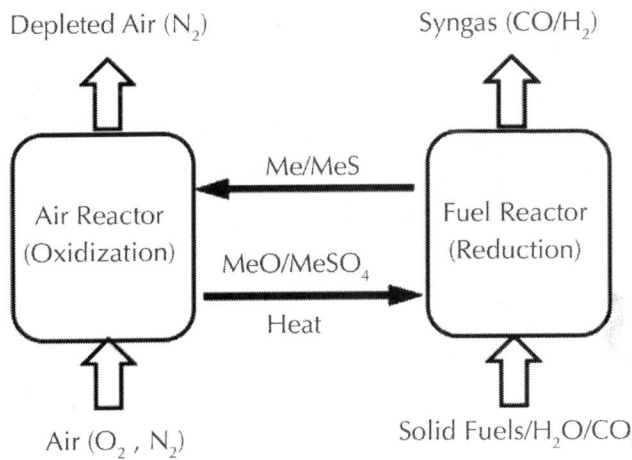

Depleted Air (N_2)

Syngas (CO/H_2)

Me/MeS

Air Reactor
(Oxidization)

$MeO/MeSO_4$

Fuel Reactor
(Reduction)

Heat

Air (O_2 , N_2)

Solid Fuels/H_2O/CO_2

Figure 1: Schematic diagram of chemical looping gasification process.

Complex reactions involving coal, $CaSO_4$ and gasifying medium in the fuel reactor include mainly the pyrolysis and gasification [Eqs. (1), (2) and (3)] of coal with CO_2/steam as the gasifying medium, reactions [Eqs.(5), (6), (7), (8) and (9)] between carbon/generated syngas and the oxygen carrier, and side reactions [Eqs. (10), (11), (12), (13), (14) and (15)] of the oxygen carrier, listed as follows.

$$C_nH_{2m}O_x \rightarrow tar + char + mixture\ gas\ (CO,\ H_2,\ CO_2,\ CH_4,\ and\ C_nH_m) \quad (1)$$

$$C + CO_2(g) \rightarrow 2CO \quad \Delta H_{298.15} = 172.423\ kJ \cdot mol^{-1} \quad (2)$$

$$C + H_2O(g) \rightarrow H_2(g) + CO(g) \quad \Delta H_{298.15} = 131.285\ kJ \cdot mol^{-1} \quad (3)$$

$$CO(g) + H_2O(g) \rightarrow H_2(g) + CO_2(g) \quad \Delta H_{298.15} = -41.138\ kJ \cdot mol^{-1} \quad (4)$$

$$CaSO_4 + 4C \rightarrow CaS + 4CO(g) \quad \Delta H_{298.15} = 520.457\ kJ \cdot mol^{-1} \quad (5)$$

$$CaSO_4 + 2C \rightarrow CaS + 2CO_2(g) \quad \Delta H_{298.15} = 175.612\ kJ \cdot mol^{-1} \quad (6)$$

$$4H_2(g) + CaSO_4 \rightarrow CaS + 4H_2O(g)$$
$$\Delta H_{298.15} = 24.51 \text{ kJ} \cdot \text{mol}^{-1} \tag{7}$$

$$4(g) + CaSO_4 \rightarrow CaS + 4CO_2(g)$$
$$\Delta H_{298.15} = -42.308 \text{ kJ} \cdot \text{mol}^{-1} \tag{8}$$

$$CH_4(g) + CaSO_4 \rightarrow CaS + 2H_2O(g) + CO_2(g)$$
$$\Delta H_{298.15} = 160.065 \text{ kJ} \cdot \text{mol}^{-1} \tag{9}$$

$$CO(g) + CaSO_4 \rightarrow CaO + CO_2(g) + SO_2(g)$$
$$\Delta H_{298.15} = 222.926 \text{ kJ} \cdot \text{mol}^{-1} \tag{10}$$

$$4CO(g) + CaSO_4 \rightarrow CaO + 3CO_2(g) + COS(g)$$
$$\Delta H_{298.15} = 77.348 \text{ kJ} \cdot \text{mol}^{-1} \tag{11}$$

$$H_2(g) + CaSO_4 \rightarrow CaO + H_2O(g) + SO_2(g)$$
$$\Delta H_{298.15} = 264.20 \text{ kJ} \cdot \text{mol}^{-1} \tag{12}$$

$$4H_2(g) + CaSO_4 \rightarrow CaO + 3H_2O(g) + H_2S(g)$$
$$\Delta H_{298.15} = 57.58 \text{ kJ} \cdot \text{mol}^{-1} \tag{13}$$

$$CaS + 3CaSO_4 \rightarrow 4CaO + 4SO_2(g)$$
$$\Delta H_{298.15} = 1060.935 \text{ kJ} \cdot \text{mol}^{-1} \tag{14}$$

$$CaSO_4 \rightarrow CaO + SO_2(g) + 1/2O_2$$
$$\Delta H_{298.15} = 50.30 \text{ kJ} \cdot \text{mol}^{-1} \tag{15}$$

In this paper, the effect of the gasifying medium on reaction characteristics between $CaSO_4$ and coal is determined based on experiments and thermodynamic analysis. The reaction mechanisms in the presence or absence of the gasifying medium are explored.

EXPERIMENTAL

Materials

The oxygen carrier used in the experiments and analysis was the pure analytical $CaSO_4$ with an average diameter of 10 μm (Tianjin Basf Chemical Co., Ltd). The proximate and ultimate analysis (air dry basis) of the coal used, Shenmu coal, is listed in Table 1. The coal samples were ground under atmospheric conditions, sieved into a 50–150 µm particle size fraction, and stored in a sealed bag. Experimental samples (200 g) were obtained by mixing $CaSO_4$ and coal with $CaSO_4$-to-coal mass ratio 2.3:1, determined by the ratio of oxygen contained in $CaSO_4$ to oxygen needed for complete conversion of Shenmu coal, represented as $C_{54.2}H_{38.3}S_{0.52}O_{6.32}N_{0.63} \cdot (H_2O)_{4.6}$ according to the proximate and ultimate analysis.

Table 1: Proximate and ultimate analysis of Shenmu coal (air dry basis)

Proximate analysis (by mass)/%				Ultimate analysis (by mass)/%					LHV/ MJ·kg^{-1}
Moisture	Volatiles	Fixed carbon	Ash	C	H	O	S	N	
8.30	29.46	52.03	10.21	65.00	3.83	11.38	0.40	0.88	24.58

LHV is the lower heat value.

Experimental Procedure

Approximately 12 mg per sample was used in the thermal gravimetric analyzer (TGA), Netzsch STA 409 PC, and Germany) to eliminate the potential effect of mass transfer between gas and solid phases. Samples were heated directly from ambient temperature to 1100 °C at a heating rate of 20 °C·min^{-1} in three gaseous media. Four additional heating rates, 5, 10, 15, and 25 °C·min^{-1}, were used to calculate the activation energy values at different degrees of conversion using an isoconversional method. The flow rates of argon and CO_2 were determined to be approximately 20 and 10 ml·min^{-1}, respectively, while that of H_2O was determined by its saturated vapor pressure in inert argon gas at 30 °C.

The gaseous products released from the TGA could not be precisely analyzed due to small amount of samples. Similar experiments were carried out to measure the effect of the gasifying medium on generated gas in a fixed-bed reactor (18Cr–12Ni–2.5Mo material type) of 55 mm in diameter and 1000 mm in height. As shown in Fig. 2, after 50 g of sample is loaded in the reactor, the gaseous medium flows into the reactor through a flow meter, passes through a steam heater, and enters a gas–solid cyclone separator. The gaseous products are then condensed *via* a condenser, metered by a gas flow meter, and finally collected using a gas collection bag for subsequent direct analysis by a gas chromatograph (PE Clarus 500). The reactor is heated directly from ambient to 900 °C at a heating rate of 20 °C·min^{-1} in three different gaseous media. The temperature is then held at 900 °C for 40 min. At 800 °C, the experiment begins to be timed and the gas product is collected. The flow rates of argon and CO_2 are determined to be approximately 40 and 20 ml·min^{-1}, respectively, while that of H_2O is determined by its saturated vapor pressure in inert argon gas at 30 °C.

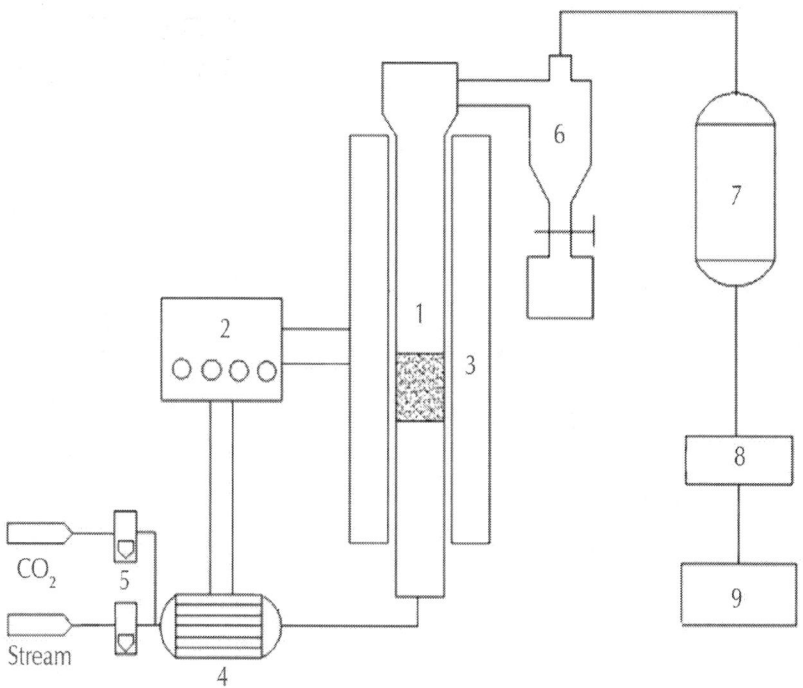

Figure 2: Schematic layout of the fixed-bed reactor setup. 1—reactor; 2—temperature control system; 3—electric heater; 4—preheater; 5—flowmeter; 6—cyclone separator; 7—condenser; 8—gas flow meter; 9—gas chromatograph analyzer.

The generation rate γ_i of syngas is calculated from

$$\gamma_i = \frac{V_{out,t} \cdot C_i}{\Delta t \cdot F_{C,\,fuel}} \quad (i = CO, CH_4, H_2, CO_2)$$

(16)

Where $V_{out,}t$ represents the volume of gaseous product in time interval Δt, C_i represents the average volumetric concentration of gaseous product i in collection bags in Δt, and $F_{C,fuel}$ is the total volume of carbon introduced to the reactor under normal conditions.

Thermodynamic Analysis

The effect of the gasifying medium on coal CLG was investigated thermodynamically based on the Gibbs free energy minimization method. The Gibbs reactor unit in Aspen software was used to simulate the coal–$CaSO_4$ system. Common gaseous species defined in the product stream are H_2 (g), H_2O (g), CO (g), CO_2 (g), and CH_4 (g); the defined sulfur species are H_2S (g), SO_2 (g), and COS (g); and the solid species are primarily CaS, $CaSO_4$, CaO, and $CaCO_3$. 1 kg coal is fed in the feeding stream, while the corresponding $CaSO_4$ and gasifying agents CO_2/steam are 13 and 54.2 mol, respectively, according to the reactions [Eqs. (5), (2) and (3)]. The carbon in the reaction is determined by the fixed carbon and carbon element in the chemical formula of coal, respectively.

RESULTS AND DISCUSSION

Thermodynamic Analysis

The variation of gaseous composition with temperature is illustrated in Fig. 3. Without the gasifying medium, the amount of CO increases while that of CO_2 decreases with the increase of temperature. Since no extra hydrogen is brought into the system, hydrogen is kept at a low level. The amount of CH_4 is seldom observed in the whole temperature range. Thus the composition of syngas comes from CO mainly. When the temperature is higher than 950 °C, the volumetric concentration of syngas (e.g. CO, H_2 and CH_4) increases slowly with temperature. When sufficient steam is brought into the system with CO_2 as the gasifying medium, both the maximum of syngas and the amount of hydrogen are higher than that without the gasifying medium. The amount of CO_2 increases with temperature, mainly due to the water shift reaction [Eq. (4)]. the concentration of syngas increases slightly at temperatures higher than 850 °C. However, the maximum concentration of syngas is just 58% because of the excessive steam gasifying medium. Similar reason is held with steam atmosphere, the maximum concentration of syngas is also 58% with H_2O as the gasifying medium. The volumetric concentration of syngas increases slowly with temperature when temperature is higher than 900 °C. Considering the three cases, we can conclude that the addition of steam and CO_2 as the gasifying medium into the system can reduce the reaction temperature, with the concentration of syngas reaching its maximum value. Under all the conditions, most of $CaSO_4$ (> 95%, by mass) is transformed to CaS. In addition, the optimal mass ratio of $CaSO_4$ oxygen carrier, coal and gasifying medium is determined by the auto-thermal balance of system, which can be referred to our previous work [13].

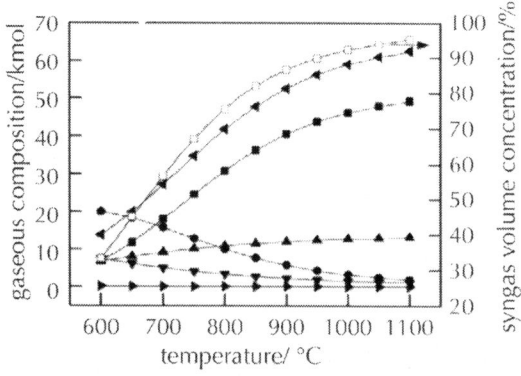

(a) Without gasifying medium

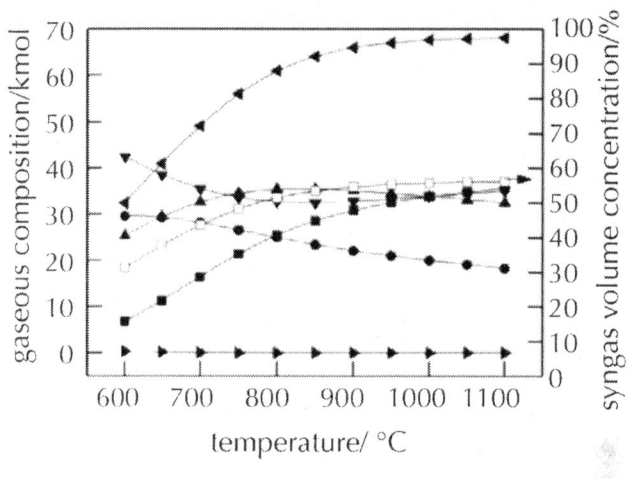

(b) With CO_2 as gasifying medium

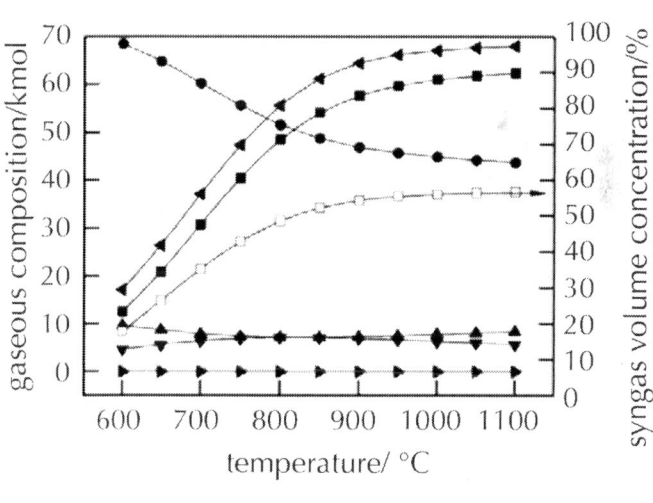

(c) With H_2O as gasifying medium

Figure 3: Variation of gaseous composition with temperature in CLG. ■ CO; CO_2; ▲ H_2; ▼ H_2O; ▶ CH_4; ◀ syngas; syngas volume concentration.

Thermal Gravimetric Experiments

Based on the thermodynamic analysis above, experiments in the presence or absence of the gasifying medium were carried out in TGA. The mass loss with temperature is indicated in Fig. 4. From the mass loss curve, we conclude that the mixture sample goes through three stages with the increase of temperature in three gaseous media: drying stage, pyrolysis stage, and chemical looping gasification stage. At temperatures lower than 400 °C, the main reaction is the drying of coal and oxygen carrier. When the temperature is between 400 °C and 850 °C, the main chemical reaction is the pyrolysis of coal. When the temperature rises above 850 °C, complex reactions between coal and $CaSO_4$ occur. However, with either CO_2 or steam as the gasifying medium, the initial temperature of complex reactions is slightly lower than that of the reactions without the gasifying medium, indicating that CO_2 and steam boost the reactions between coal and oxygen carrier. The mass loss rate of the mixture sample between 750 °C and 1100 °C is also shown in Fig. 4. From the derivative thermogravimetric analysis (DTG) curve, it is evident that with steam as the gasifying medium, the maximum mass loss rate occurs at 940 °C, while that without the gasifying medium or with CO_2 as the medium occurs at 980 °C. The maximum mass loss rate is 7% per minute. The derivative thermogravimetric analysis (DTG) curve illustrates that steam promotes the reaction between coal and oxygen carrier by reducing the reaction temperature. Moreover, the mass loss rate with the gasifying medium (steam or CO_2) is greater than that observed in systems without the gasifying medium at the same temperature below 950 °C.

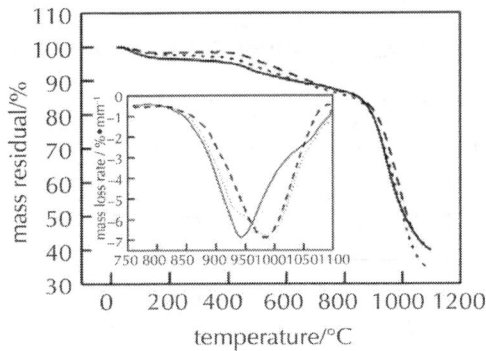

Figure 4: TG and DTG curves as a function of temperature in the presence or absence of a gasifying medium. — Steam; -- N_2; ···· CO_2.

The DTG curves in Fig. 4 may be the superposition of several mass loss rate peaks of different reactions at different temperatures at temperatures beyond 750 °C. Hence, a peak-fitting method utilizing a Gaussian distribution as the fitting function is used to analyze the complex DTG curves. The fitting peaks and individual peaks are displayed in Fig. 5. In the absence of the gasifying medium, the DTG curve comprises two individual peaks at 950 °C and 1000 °C; with CO_2 as the gasifying medium, two individual peaks are observed at 910 °C and 980 °C; and with steam as the gasifying medium, three individual peaks occur at 875 °C, 940 °C and 1025 °C. These differences can be explained by the three different environments with complex reactions of the coal-$CaSO_4$ system. In the inert atmosphere, the first peak is due to Eq. (5) or Eq. (6) between coal char and $CaSO_4$ that produce CO or CO_2; the second peak at 1000 °C is formed by Eq. (8) or the gasification reaction [Eq. (2)] of coal char. With CO_2 as the gasifying medium, the first peak begins at 840 °C and the following peak starts at 850 °C, resulting from gasification reaction Eq. (2) of coal char, which further promotes the reaction between coal char and $CaSO_4$ through Eq. (8). Using steam as the gasifying medium, because the gasification temperature of coal char is lower than that of CO_2 medium, the gasification reaction [Eq. (3)] (which generates syngas) begins at 800 °C, resulting in the first peak. The generated syngas further promotes the reaction between coal char and $CaSO_4$ through Eqs. (7) And (8), as indicated by the second peak beginning at 850 °C. The third peak at 950 °C is attributed to the reaction between coal char and $CaSO_4$ because of the effect of diffusion, which will be discussed later.

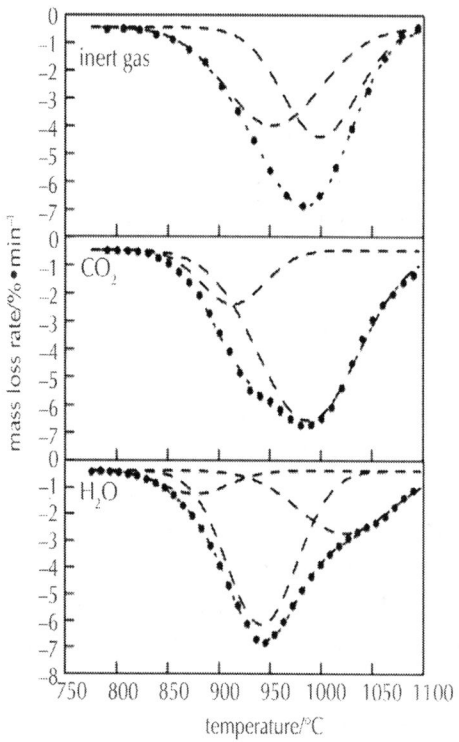

Figure 5: Peak fitting of DTG curves in the presence or absence of a gasifying medium.   Original data; -- individual peaks; ⋯⋯ fitting line.

The comparison of fitting curves in each of the three gasification conditions gives the effect of the gasifying medium on CLG. Both CO_2 and steam promote the reactions between coal char and $CaSO_4$. In a CO_2 atmosphere, the gasification of coal char and the reaction between coal char and $CaSO_4$ start at approximately the same temperature. However, the gasification of coal char in steam occurs at a lower temperature, with the reaction between coal char and $CaSO_4$ divided into two evident stages. The main reason for the difference between these two gasifying media is related to their reactivity. The reaction rate of $CaSO_4$ with H_2 is different from that with CO.

Effect of Gasifying Medium on Generation Rate

To understand the effect of the gasifying medium on the generation rate of syngas, experiments similar to the TG experiments were carried out. The generation rate of syngas in the fixed-bed reactor as a function of temperature and time under three conditions is shown in Fig. 6. The gas generation rates of CH_4, H_2 and CO reach their maximum values as the temperature rises to 800 °C, while that of CO_2 continues to increase. The gas generation rates of CO_2 and CO increase at 60 min. The generation of CH_4, H_2 and CO at temperatures less than 900 °C mainly originates from the pyrolysis of coal, while at 900 °C the production of CH_4, H_2 and CO originates from the reactions between coal and $CaSO_4$. With CO_2 as the gasifying medium, CH_4 and H_2 are mainly generated from the pyrolysis of coal. However, the generation rate of CO is higher than that of other gases. CO is primarily generated from the gasification of coal at temperatures less than 900 °C; the sudden increase in CO production at 900 °C is due to the gasification reaction and the reaction between coal and oxygen carrier. With steam as the gasifying medium, CO_2 and H_2 are produced at higher rates. At 900 °C, the increased production rates of CO, CO_2 and H_2 are mainly due to the reaction between coal and oxygen carrier as well as the water shift reaction. The generation rate of pyrolysis gases is mostly the same under three conditions. However, the generation rates of CO with CO_2 as the gasifying medium and H_2 with steam as the gasifying medium are highly elevated. The gasification reaction of coal can improve the generation rate and the reaction between coal and oxygen carrier. Using either CO_2 or steam as the gasifying medium can promote the reactions between coal and oxygen carrier to generate more syngas.

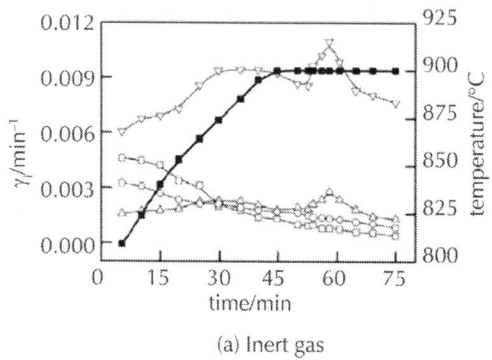

(a) Inert gas

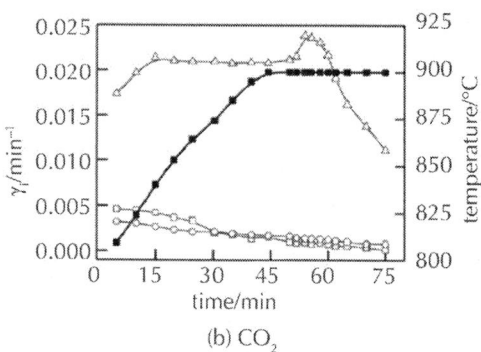

(b) CO_2

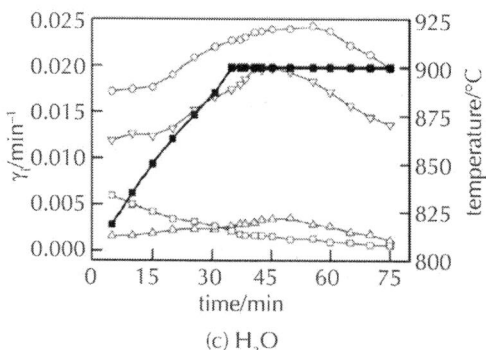

(c) H_2O

Figure 6: Gas generation rate under three conditions: (a) inert gas; (b) CO_2; (c) H_2O. △ CO; CH_4; ▽ CO_2; H_2; ■ temperature.

Reaction Mechanisms under Different Atmospheres

To investigate the reaction mechanism of syngas production from the chemical looping gasification process, thermogravimetric analyses for the experimental samples were carried out at five heating rates under three conditions. The Flynn–Wall–Ozawa isoconversional method was used in this study. A description of the method can be found in our previous publication [14].

The activation energy as a function of conversion degree α is presented in Fig. 7. In the absence of the gasifying medium, the activation energy increases with α up to 0.42 and subsequently decreases. Compared to an inert atmosphere, the activation energy with either CO_2 or steam as the gasifying medium is small, indicating that CO_2 and steam can improve the reaction by reducing the reaction activation energy. However, because of the complex reactions in the system, the activation energy as a function of conversion degree is a curve. With CO_2 as the gasifying medium, the activation energy decreases with the increase of conversion degree up to 0.65 and increases thereafter, with the minimum activation energy of 248.7 $kJ \cdot mol^{-1}$. With steam as the gasifying medium, the variation of the activation energy is the same as that without the gasifying medium. However, the maximum activation energy with steam as the gasifying medium is 228.6 $kJ \cdot mol^{-1}$ at $\alpha = 0.55$; in the absence of the gasifying medium, the maximum activation energy is 413.8 $kJ \cdot mol^{-1}$ at $\alpha = 0.42$. Based on these observations, steam can boost the reaction between coal char and oxygen carrier.

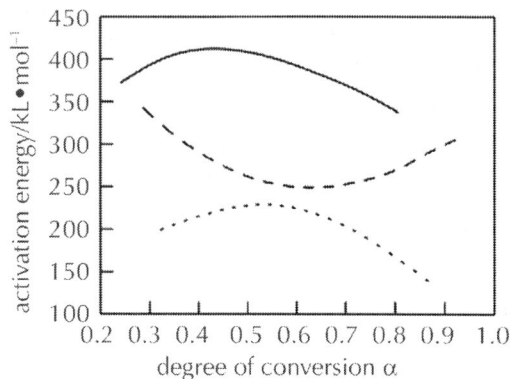

Figure 7: Activation energy with the degree of conversion under three conditions. — N_2;-- CO_2; ⋯⋯ H_2O.

Vyazovkin and Lesnikovich [15] and Dowdy [16] have presented that a complex process/reaction is identified with variable activation energies at different   values while the activation energy of a single stage reaction is kept constant with   changing for the isoconversional method. They have shown that the activation energies of simultaneous/parallel reactions, during which at least two reactions occur simultaneously, are increasingly dependent on the degree of conversion, while decreasing dependency is typical of complex reactions with potential alternative limiting stages; the limiting stage could be a kinetic control stage or a diffusion control stage. Based on this observation, the reaction mechanism can be obtained from Fig. 7. The complex reactions in the beginning stage of reactions without the gasifying medium and reactions with steam as the gasifying medium are simultaneous or parallel reactions. Reactions in the following stage are complex reactions with the limiting stage occurring in a diffusion regime instead of a kinetic regime. These two different stages can account for the existence of two reaction peaks between coal char and $CaSO_4$. With CO_2 as the gasifying medium, the reaction stages are reversed. A change in the limiting stage from a kinetic to a diffusion regime occurs first, followed by the simultaneous/parallel reactions. This result can be observed in the SEM images of reacted residues of experimental sample at 1100 °C in TGA. As shown in Fig. 8, the reacted residue with CO_2 as the gasifying medium is loose and more porous than the residue obtained with steam as the gasifying medium. The sintering phenomenon of the residue in the H_2O atmosphere prevents further reaction and the diffusion stage limits the complex reaction.

(a) With CO_2 as the gasifying medium

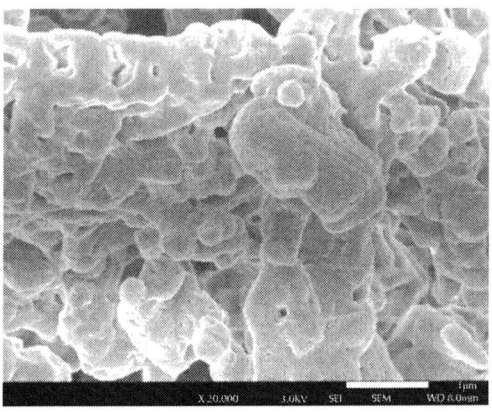

(b) With H_2O as the gasifying medium

Figure 8: SEM images of reacted residues with CO_2 (a) or H_2O (b) as the gasifying medium.

CONCLUSIONS

Based on thermodynamic analysis and experiments in TGA and fixed-bed reactor, the effect of the gasifying medium on the reactions between $CaSO_4$ and coal and the syngas generation rate is examined. The following conclusions are obtained.

1. The addition of steam and CO_2 as the gasifying medium in the system can reduce the temperature from 950 °C without the gasifying medium to 850 °C and 900 °C, respectively, at which the volumetric concentration of syngas approaches its maximum value.

2. Thermal gravimetric analysis indicates that the mixture sample goes through three stages in three gaseous media: drying stage, pyrolysis stage and chemical looping gasification stage as temperature increases from 800 °C to 1100 °C.

3. Both steam and CO_2 as the gasifying medium can promote the CLG for syngas. The maximum activation energy reduces from 413.8 $kJ \cdot mol^{-1}$ without the gasifying medium to 228.6 $kJ \cdot mol^{-1}$ with steam as the gasifying medium. With steam as the gasifying medium, parallel reactions occur in the beginning stage, followed by a limiting stage in the diffusion regime, which is opposite to the reaction mechanism with CO_2 as the gasifying medium.

REFERENCES

1. J. Adanez, A. Abad, F. Garcia-Labiano, P. Gayan, L.F. de Diego, Progress in chemical looping combustion and reforming technologies, Prog. Energy Combust. 38 (2) (2012) 215–282.

2. J.S. Wang, E.J. Anthony, Clean combustion of solid fuels, Appl. Energy 85 (2–3) (2008) 73–79.

3. T. Mattisson, A. Lyngfelt, H. Leion, Chemical-looping with oxygen uncoupling for combustion of solid fuels, Int. J. Greenh. Gas Control 3 (1) (2009) 11–19.

4.   L.S. Fan, F.X. Li, S. Ramkumar, Utilization of chemical looping strategy in coal gasifi- cation processes, Particuology 6 (3) (2008) 131–142.

5.   P. Chiesa, G. Lozza, A. Malandrino, M. Romano, V. Piccolo, Three-reactor chemical looping process for hydrogen production, Int. J. Hydrog. Energy 33 (9) (2008) 2233–2245.

6.   M. Najera, R. Solunke, T. Gardner, G. Veser, Carbon capture and utilization via chemical looping dry reforming, Chem. Eng. Res. Des. 89 (9A) (2011) 1533–1543.

7.   T. Mattison, A. Lyngfelt, Applications of chemical looping combustion with capture of CO2, Proc. of the 2nd Nordic Minisymposium on Carbon Dioxide Capture and Storage, Göteborg, Sweden, 2001.

8.   F. He, Y.G. Wei, H.B. Li, H. Wang, Synthesis gas generation by chemical-looping reforming using Ce-based oxygen carriers modified with Fe, Cu, and Mn oxides, Energy Fuel 23 (2009) 2095–2102.

9.   L.F. de Diego, M. Ortiz, J. Adanez, F. Garcia-Labiano, A. Abad, P. Gayan, Synthesis gas generation by chemical-looping reforming in a batch fluidized bed reactor using Ni-based oxygen carriers, Chem. Eng. J. 144 (2) (2008) 289–298.

10.   L.F. de Diego, M. Ortiz, F. Garcia-Labiano, J. Adanez, A. Abad, P. Gayan, Hydrogen production by chemical-looping reforming in a circulating fluidized bed reactor using Ni-based oxygen carriers, J. Power Sources 192 (1) (2009) 27–34.

11.   H.E. Andrus, J.H. Chiu, P.R. Thibeault, A. Brautsch, Alstom's calcium oxide chemical looping combustion coal power technology development, Proc. of 34th International Technical Conference on Clean Coal & Fuel Systems, Florida, USA, 2009.

12.   M. Zheng, L.H. Shen, J. Xiao, Reduction of CaSO4 oxygen carrier with coal in chemical-looping combustion: effects of temperature and gasification intermediate, Int. J. Greenh. Gas Control 4 (5) (2010) 716–728.

13. Y.Z. Liu, Q.J. Guo, Investigation into syngas generation from solid fuel using CaSO4- based chemical looping gasification process, Chin. J. Chem. Eng. 21 (2) (2013) 127–134.

14. Y.Z. Liu, Q.J. Guo, Y. Cheng, H.J. Ryu, Reaction mechanism of coal chemical looping process for syngas production with CaSO4 oxygen carrier in the CO2 atmosphere, Ind. Eng. Chem. Res. 51 (31) (2012) 10364–10373.

15. S.V. Vyazovkin, A.I. Lesnikovich, An approach to the solution of the inverse kinetic problem in the case of complex processes: Part I. Methods employing a series of thermoanalytical curves, Thermochim. Acta 165 (2) (1990) 273–280.

16. D.R. Dowdy, Meaningful activation energies for complex systems: The application of Ozawa–Flynn–Wall method to multiple reactions, J. Therm. Anal. Calorim. 32 (1987) 137–147.

Cao-based Chemical Looping Gasification of Biomass for Hydrogen-enriched Gas Production with in Situ CO$_2$ Capture and Tar Reduction

Jakkapong Udomsirichakorn[a], Prabir Basu[b], P. Abdul Salam[a], and Bishnu Acharya[c]

[a]Energy Field of Study, Asian Institute of Technology, Khlong Luang, Pathumthani 12120, Thailand

[b]Mechanical Engineering Department, Dalhousie University, Halifax B3J 1Z1, Canada

[c]Greenfield Research Incorporated, Halifax B3M 1N8, Canada

ABSTRACT

Steam gasification of biomass undergoes the problem of undesirable CO_2 and tar formation. Calcium oxide (CaO), when added to the gasification, could play the dual role of tar reforming catalyst and CO_2 sorbent, and thereby produce more hydrogen. However, the deactivation of CaO after carbonation reaction is challenging for continuous hydrogen production and economical perspective. The concept of CaO-based chemical looping gasification (CaO-CLG) plays a key role in overcoming such a challenge. This work primarily aims at studying steam gasification of biomass with the presence of CaO in a uniquely designed chemical looping gasification (CLG) system for hydrogen production with in situ CO_2 capture and tar reduction. The effect of solid circulation rates on gas and tar production is studied. A comparison of CaO-CLG, sand-based chemical looping gasification (Sand-CLG) and CaO-based bubbling fluidized bed gasification (CaO-BFBG) is presented mainly focusing on gas and tar production. The maximum H_2 and minimum CO_2 concentrations as well as maximum H_2 yields of 78%, 4.98% and 451.11 ml (STP)/g of biomass, respectively, are obtained at the solid circulation rate of 1.04 kg/m²s. At the same point, the maximum total gas yield was 578.38 ml (STP)/g of biomass and the tar content of 2.48 g/Nm³ was the lowest. 30% higher concentration of H_2 and triple yield of H_2 were found in CaO-CLG compared to Sand-CLG. Compared to CaO-BFBG, CaO-CLG resulted in 15% higher concentration of H_2 and almost double yield of H_2. Moreover, the lowest tar content of 2.48 g/Nm³ was obtained for CaO-CLG while the tar content was 68.5 g/Nm³ for Sand-CLG and 26.71 g/Nm³ for CaO-BFBG. CO_2 concentration obtained for CaO-CLG also significantly reduced by 13–17% as compared to both Sand-CLG and CaO-BFBG.

INTRODUCTION

The continued use of fossil based energy is increasingly contributing to global warming, climate change and energy security issues. As a consequence, many researches and developments worldwide are realized to encourage the use of renewable and sustainable energy sources. Hydrogen being a pollution-free energy carrier is expected to be the most promising source to replace fossil fuel employed both in power generation and transportation sectors [1], [2] and [3]. However, it is not naturally available in sufficient quantities and also needs to be synthetically produced. Currently, there are a number of energy sources and technologies to produce hydrogen. But about 96% of the hydrogen for commercial use is produced from fossil fuels, nearly 50% of which is contributed by natural gas, primarily via steam methane reforming [4], which is a fossil fuel based process. Biomass is considered as the potential substitute for the depleting fossil fuels [5]. It is also accepted as the greatest promise due to its availability everywhere in the world [2].

The technologies available for conversion of biomass into hydrogen-rich gas can be classified into biological and thermo-chemical methods [6], [7], [8], [9] and [10]. Biomass steam gasification, as one of the thermo-chemical methods, has been perceived as an attractive process for producing syngas rich in hydrogen [11], [12], [13], [14], [15] and [16]. However, the process unavoidably suffers from the problem of unpleasant tar and CO_2 formed within the process. The use of calcium oxide (CaO) dually acknowledged as a tar reforming catalyst and a CO_2 sorbent has currently gained lots of attention due to its cheapness and abundance [17]. Its role in catalytic reforming of tar not only reduces the tar amount in the product gas but also enhances the total gas and hydrogen yields [1], [17] and [18]. Similarly, another role in removing CO_2 from the gasification reaction as soon as it is formed alters the equilibrium composition of the produced gas and promotes the production of gas rich in hydrogen [3] and [17]. However, the deactivation of CaO after capturing CO_2 seems to be a major obstacle to continuous hydrogen production. Also, if the frequent replacement of CaO sorbent is needed, the process may not be economically attractive [19]. To overcome such challenges, the concept of CaO-based chemical looping gasification (CaO-CLG), basically aimed for hydrogen production with in situ CO_2 capture, is playing an important role.

The CaO-CLG concept was initiated through the CO_2 acceptor process and has currently been developed through the HyPr-RING process, the ZECA process, the ALSTOM process, and the AGC process [20]. Nevertheless, some of these processes are not considered as environmentally sustainable hydrogen production processes due to their dependence on fossil fuel. Moreover, the processes were designed to operate at high pressures and/or temperatures and some of them were designed with multiple-loop configuration that causes high operational complexity as well as high capital cost. Although these demerits are challenging for the operation of chemical looping process for continuous hydrogen production, no formation of CO_2 and tar observed in some previous processes, such as the HyPr-RING, seems to be advantageous. This shows that each technology has its own pros and cons. To facilitate such challenge of high operational complexity of the past technologies and encourage environment-friendly hydrogen production, recent efforts [21] and [22] have been devoted to developing the simpler CaO-CLG system with the single-loop and atmospheric operation as well as utilization of biomass and biowaste. A real CaO-CLG system which mainly consists of regenerator, gasifier and loopseal as shown in Fig. 1 was uniquely designed and developed by Acharya et al. [21] and [22] based on the mature circulating fluidized bed (CFB) technology and was experimentally used in the present work as well. Acharya et al. [21] studied the influence of temperature and in-bed CaO on hydrogen and CO_2 production in a gasifier and also the regeneration of calcium carbonate ($CaCO_3$) in a regenerator. Recently, Acharya et al. [22] tested the influence of three different calcination media that are necessary and used for supplying to a regenerator, whereas Udomsirichakorn et al. [17] investigated the influence of temperature and steam-to-biomass ratio (S/B) as well as in-bed CaO on syngas and tar produced from a gasifier. Moreover, the capability of the CaO sorbent as a bed material undergoing calcination–carbonation cycles is also studied by Acharya et al.[22]. These previous experimental studies have been conducted using the same CaO-CLG system but separately focused on individual component and sub-process of the system. Overall investigation on the whole looping system is still lacking not only in this CaO-CLG system but also in other CaO-CLG systems that are scarcely found in open literature. This provides an opportunity for the present study to fulfill the knowledge gap on the CaO-CLG system which is currently available only in R&D stage [23].

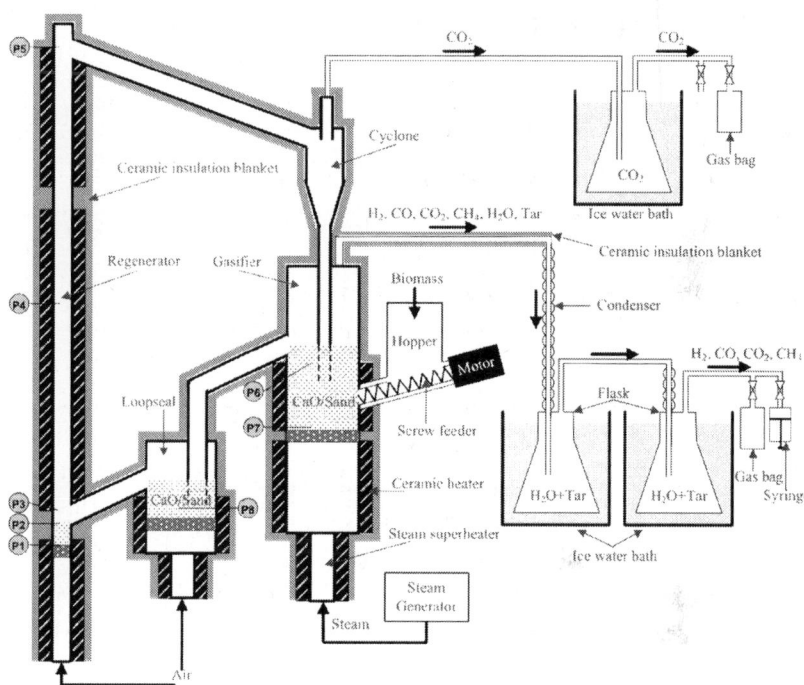

Figure 1: Experimental setup.

This paper presents an experimental study of catalytic steam gasification of biomass in the CaO-CLG system with the main focus of studying the influence of solid circulation rate on the combined role of CaO as a tar reforming catalyst and CO_2 sorbent to enhance hydrogen production. The comparative study of the production of syngas and tar obtained from CaO-CLG and sand-based chemical looping gasification (Sand-CLG) as well as CaO-based bubbling fluidized bed gasification (CaO-BFBG) is also presented here.

EXPERIMENTAL

Biomass Feedstock

Pine wood sawdust with particle size of 0.425–0.5 mm was used as the feedstock. The proximate analysis of the sawdust was conducted following ASTM standard test methods. The higher heating value (HHV) analysis was performed according to the British Standard No. BS4379 using a bomb calorimeter (Parr 6100). The carbon, hydrogen, nitrogen and sulfur contents (C, H, N and S) in the sawdust were determined using an elemental analyzer (Leco CHN-1000 and Leco SC-432) and oxygen (O) was then calculated by difference to 100%. The ultimate and proximate analysis results are presented in Table 1.

Table 1: Ultimate and proximate analysis of biomass

Feed-stock	Ultimate analysis (wt.%, dry ash-free basis)					HHV (MJ/kg)	Proximate analysis (wt.%, as-received basis)			
	C	H	N	S	O		M	VM	Ash	FC
Pine sawdust	53.28	5.55	0.01	0.003	41.15	18.70	6.16	85.42	0.53	7.89

Bed Materials

Silica sand (mean particle size 0.25–0.3 mm) with apparent density of 2300 kg/m^3 and calcined limestone (mean particle size 0.25–0.3 mm) were used as bed materials for the Sand/CaO-CLG system. Prior to starting each experiment, the sized limestone (apparent density 2600 kg/m^3) was calcined in a muffle oven at 900 °C for 3 h and then gradually cooled down by purging the atmosphere in the oven with nitrogen gas to avoid a carbonation reaction of calcined limestone. The chemical composition of calcined limestone is presented in Table 2. The change in particle size of limestone after calcination was found to be very small, but its weight loss through calcination was about 42.7%.

Table 2: Chemical composition of calcined limestone in wt.%

CaO	MgO	SiO$_2$	Fe$_2$O$_3$	Al$_2$O$_3$	CaCO$_3$
95.5	0.9	1.3	0.2	0.4	1.1

Facilities and Procedures

Fig. 1 shows the schematic drawing of the chemical looping gasification system developed by Acharya et al. [21] and [22] and used in this experimental study. The unique tailor-made configuration of this system mainly consists of a bubbling fluidized-bed gasifier with a height of 450 mm, a circulating fluidized-bed regenerator with a height of 1500 mm and a bubbling fluidized-bed loopseal with a height of 135 mm. Electric heaters with a total capacity of 12 kW were mounted alongside these reactors and used to externally heat the system. All experiments were carried out at atmospheric pressure. Before starting each run, about 1.5–2 kg of either fresh CaO (calcined limestone) or sand with a mean particle size of 0.275 mm was manually fed into the bed of the reactors for use as bed material. The biomass feedstock was loaded into the hopper. After loading the gasifier, all the heaters as well as the electric steam generator and steam/air superheater/preheater were turned on. For initial heating up of the system, air was supplied through a 30-μm porous ceramic distributor plate of each reactor to fluidize the bed material. Moreover, the temperature and pressure of all reactors were measured by K-type thermocouples and pressure transducers. It took around 3–4 h to reach a bed temperature of 650 °C in the gasifier, of 900 °C in the regenerator and of 350 °C in the loopseal. Once the bed attained these temperatures, the air flow through the bed of gasifier was replaced with superheated steam supplied from the electric steam generator and superheated by the electric steam superheater installed below the gasifier. But the fluidization in the regenerator and in the loopseal was kept supplying with air but increased to higher velocities.

Table 3 shows the flow rate and superficial velocity of different fluidization media supplied to each reactor based on theoretical calculation and practical considerations for suitable fluidizing regime. Unlike other reactors, the aeration provided to the loopseal for all experiments was sufficient to give a circulation of the solid from one to another reactor as a looping process and this was also varied to allow different desired solid circulation rates. Once the system was stabilized at the desired temperature in a circulating mode of the solid, the biomass feeding was started from the hopper through a water cooled screw pipe. Thus, biomass nearly at room temperature mechanically moved into the bed at the desired constant feeding rate that was regulated by the variable speed drive. Inside the system, unburned biomass char obtained from gasification leaves the gasifier together with the bed material through an inclined, gasifier-loopseal connecting the pipe towards the loopseal operated in a bubbling mode and then go through an inclined, loopseal-regenerator connecting the pipe towards the riser serving as a regenerator. Inside the riser-regenerator, the heat internally generated from char combustion was transferred to the solid and also used to regenerate calcium carbonate ($CaCO_3$) produced in and circulated from the gasifier. The air supplied to the bed in the regenerator causes a fast fluidization so that the solid gets entrained and separated in the cyclone. Once separated, the hot bed material falling through the standpipe will start filling the gasifier and simultaneously providing heat to the gasification process. Despite being heated by the circulating hot solid, the gasification process was still being heated by electric heaters to maintain the bed temperature constant at 650 °C for all experiments. Once the gasification took place steadily, that typically occurred 10–15 min after feeding the biomass, the hot syngas from the gasifier flowing through the sampling line, insulated with ceramic insulation blanket, was condensed in cold traps where the condensed steam and tar as well as solid particles were collected in flasks. The dry and clean gas then passed through the sampling point where a Tedlar gas bag was connected and the sample was taken every 5 min for about 1 h. The gas flow rate was further measured and recorded at some particular moments of the experiment. After the end of the experiment, the gas samples were analyzed in a gas chromatograph (VARIAN Micro GC-490) and the condensate collected in the flasks was taken for tar analysis. The liquid sample of tar, collected after each experimental run, was paper-filtered, homogenized and then used for quantitative analysis of gravimetric tar. The concentration of gravimetric tar was determined by evaporating the homogenized liquid sample in a standard rotary evaporator, according to the Tar Protocol CEN/TS 15439:2006 [24].

Table 3: Flow streams in different fluidizing reactors

Reactor	Fluidizing medium	Fluidization regime	Total flow rate (Nm³/s)	Fluidizing velocity (m/s)
Gasifier	Steam	Bubbling bed	7.40×10^{-4}	0.09
Loopseal				
at $G_s^a = 0.91$ kg/m²s	Air	Bubbling bed	2.33×10^{-4}	0.029
at $G_s = 1.04$ kg/m²s	Air	Bubbling bed	2.70×10^{-4}	0.033
at $G_s = 1.14$ kg/m²s	Air	Bubbling bed	3.25×10^{-4}	0.04
Regenerator[b]				
at $G_s = 0.91$ kg/m²s	Air	Fast bed	1.10×10^{-3}	2.17
at $G_s = 1.04$ kg/m²s	Air	Fast bed	1.17×10^{-3}	2.30
at $G_s = 1.14$ kg/m²s	Air	Fast bed	1.27×10^{-3}	2.51

[a]G_s = Solid circulation rate.

[b]Total flow rate of air to regenerator = Fixed flow rate of air to regenerator + Varied flow rate of air to loopseal.

RESULTS AND DISCUSSION

Hydrodynamics and Pressure Distribution

Fig. 2 shows the average pressure distribution in the present CLG system to verify the circulating fluidization regime in this CFB based looping system. This is similar to what one would expect in a CFB system [22] and [25]. The average voidage in the bubbling fluidized-bed gasifier was found to be 0.61. The average voidage in the turbulent bed section of the riser-regenerator was 0.82, which increases to 0.96 around its exit. The pressure balance around the loop shows that the pressure inside the gasifier is much higher than that at the top of the loopseal so that no air going into the loopseal could flow into the gasifier. This was also verified by the negligible amount of air composition detected in the product gas measured at the outlet of the gasifier. Moreover, the pressure at the base of the loopseal is much higher than that in the regenerator so that no gas can escape the regenerator through the loopseal.

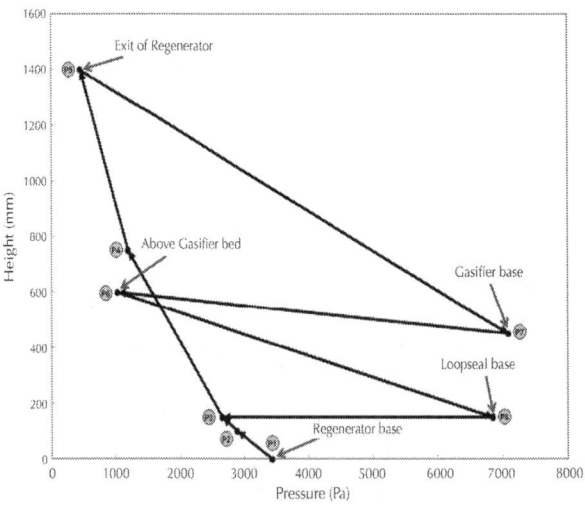

Figure 2: Pressure distribution in the CaO-CLG system.

Fig. 3 presents the average suspension density profile in the riser-regenerator calculated from the pressure measured along its height. The nature of this suspension profile is similar to the one observed in CFB boilers [25]. These two characteristics of a nature of a CFB regime found in this study can confirm that the system was operated in suitable CFB-based looping process.

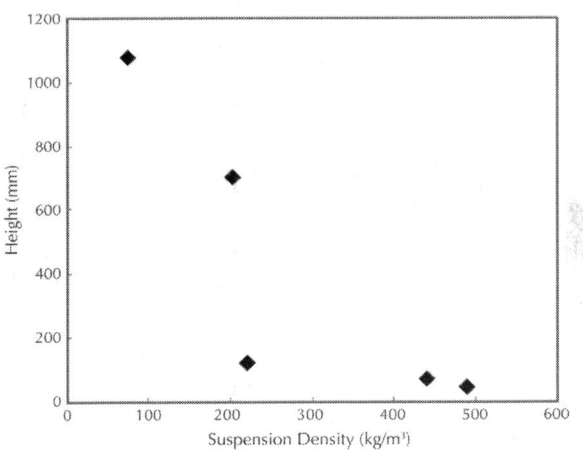

Figure 3: Suspension density profile along the height of the riser-regenerator.

Influences of Solid Circulation Rate on Production of Syngas and Tar

Fig. 4 presents the influence of solid circulation rate on product gas composition and H_2 yield obtained at identical operating conditions (S/B = 3.41 and gasification temperature = 650 °C). Fig. 5 shows its effect on the tar and product gas yields. Experimental results (Fig. 4) show the concentration and yield of H_2 increase with the increment of circulation rate from 0.91 to 1.04 kg/m²s and then decrease with further increase of circulation rate from 1.04 to 1.14 kg/m²s. The similar trend is found for the total gas yield but the opposite is observed for the tar yield (Fig. 5).

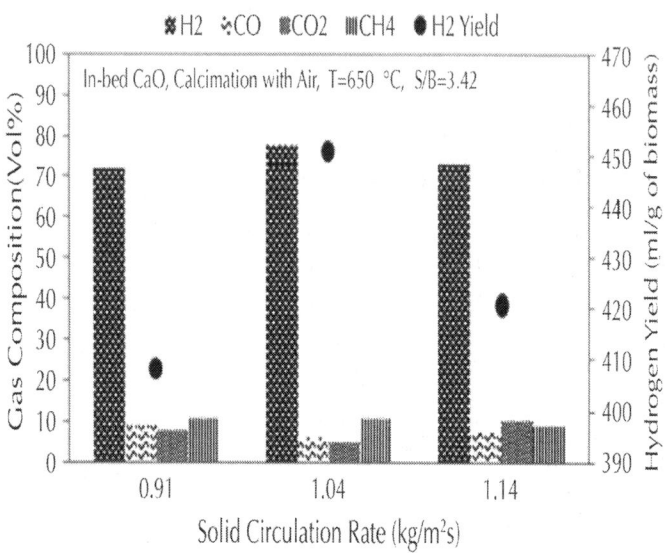

Figure 4: Effect of solid circulation rate on gas composition and hydrogen.

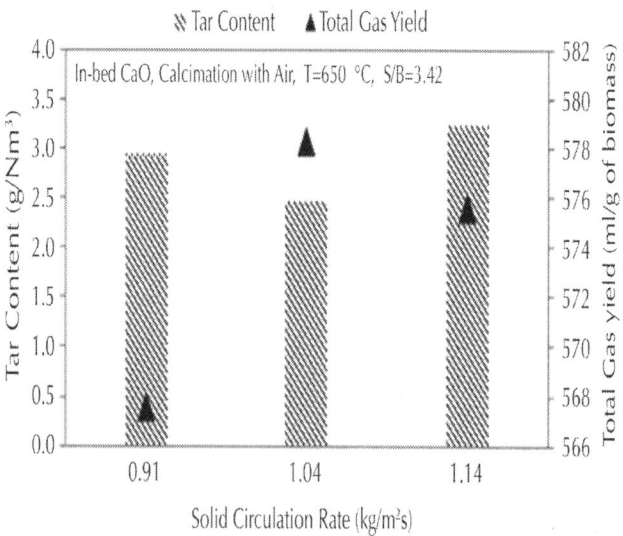

Figure 5: Effect of solid circulation rate on yields of tar and gas.

Fig. 4 shows that as the solid circulation rate increased from 0.91 to 1.04 kg/m²s the volumetric concentration of H_2 sharply increased from 71.93% to 78% and its yield also increased over 40 ml/g of biomass. The main reason for this phenomenon is probably attributed to the influence of the circulation rate on the quantity of CaO derived from regenerating $CaCO_3$ via a calcination reaction (Eq. (1)) in the regenerator and then circulated to the gasifier, in which CO_2 and tar are produced. This can be also implied that a higher circulation rate allows a higher amount of regenerated CaO in the gasifier. As a consequence, more sites are available for higher CO_2 absorption by CaO via a carbonation reaction (Eq. (2)). This conjecture was also reported by Udomsirichakorn et al. [17] and could be supported by the decrease in CO_2 concentration from 8.07% to 4.98% observed for the present study. The higher absorption of CO_2 consequently lowers CO_2 partial pressure and then enhances water–gas shift reaction (Eq. (3)) to move in forward direction to produce more H_2 [9], [26] and [27]. This promotion of water–gas shift reaction (Eq. (3)) could be also evidenced by the observed decrease of CO concentration from 9.04% to 5.93%, which is consistent with the increase of H_2 concentration.

Calcination reaction : $CaCO_3 \rightarrow CaO + CO_2$

(1)

Carbonation reaction : $CaO + CO_2 \rightarrow CaCO_3$

(2)

Water-gas shift reaction . $CO + H_2O \rightarrow CO_2 + H_2$

(3)

Catalytic tar reforming reaction : $Tars + H_2O \xrightarrow{Cao} H_2 + CO_2 + CO + hydrocarbons + ...$

(4)

Fig. 5 shows a decrease of tar content from 2.95 to 2.48 g/Nm³ with an increase of circulation rate from 0.91 to 1.04 kg/m²s. The similar trend was reported by Pfeifer et al. [28]. Moreover, the present observation of tar amount can also support the above conjecture of more availability of CaO in the gasifier due to higher circulation rate of solid. As a result of more CaO in the gasifier, the reduction of tar content can be attributed to the enhancement of catalytic reforming of tar with CaO (Eq. (4)) resulting in the consistent increase of the yields of H_2 (Fig. 4) and total gas (Fig. 5) as reaction products. This also shows that CaO can simultaneously play another role in reforming tar into gas apart from CO_2 absorption [9], [17], [26],[27] and [29]. In-depth analysis of this is well discussed elsewhere [17].

However for further increase of circulation rate beyond 1.04 kg/m²s, a sudden decrease in H_2concentration to 73.08% with consistent increase in CO_2 concentration to 10.06% was noted (Fig. 4). This could be attributed to a short residence time of $CaCO_3$ in the regenerator along with low calcination level of it, which consequently results in less amount of reactivated CaO in the gasifier. This speculation is also supported by the increase of tar content that is consistent with the decrease of total gas yield within the range of circulation rate of 1.04–1.14 kg/m²s (Fig. 5).

Overall, it can be concluded that even though a higher circulation rate of bed material allows positive effect on production of H_2, CO_2 and tar, too high circulation rate can adversely affect the calcination process of $CaCO_3$, which consequently results in negative production of syngas and tar. However, due to unavailability of this information in CaO-CLG system, more studies are encouraged to validate the present results.

Influences of Cao-CLG on Production of Syngas and Tar

The main focus of this experimental study is steam gasification of biomass in the CaO-CLG system. To point out its superiority in terms of gas and tar production, two baseline studies such as Sand-CLG as well as CaO-BFBG were conducted for comparison. The experimental results obtained for different gasification systems but at identical operating conditions (S/B = 3.41 and gasification temperature = 650 °C for all the systems, and solid circulation rate = 1.04 kg/m²s for both CLG systems) are compared in Fig. 6 for gas composition and H_2 yield and in Fig. 7 for tar and total gas yields. It is evidenced that by using CaO-CLG, one can produce the product gas with the highest concentration and yield of H$_2$ and the highest yield of total gas as well as lowest contents of CO$_2$ and tar.

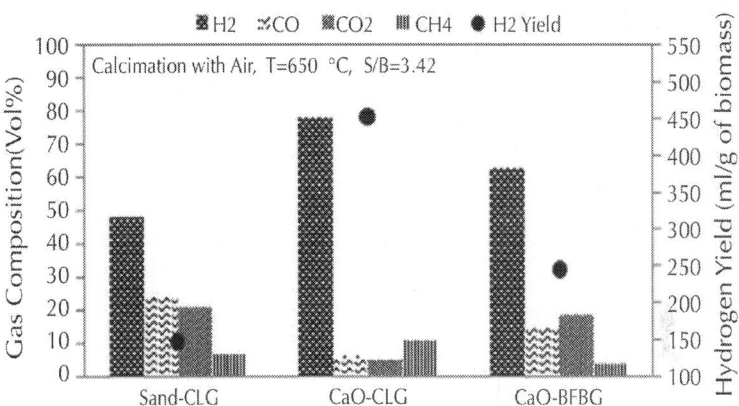

Figure 6: Effect of CaO-CLG on gas composition and hydrogen.

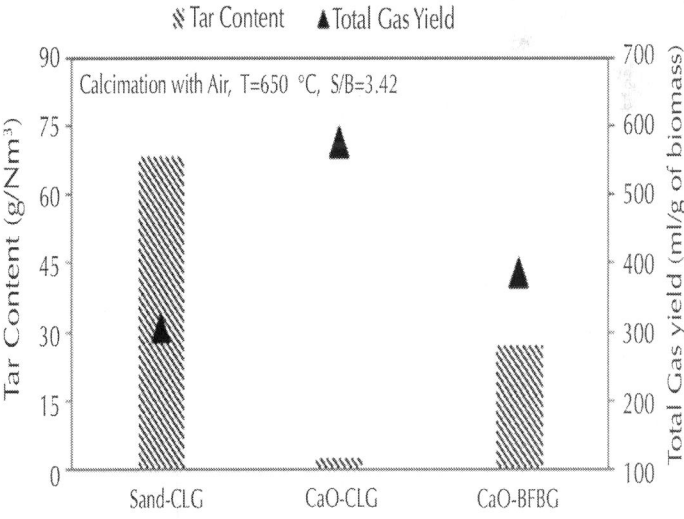

Figure 7: Effect of CaO-CLG on yields of tar and gas.

As the influence of CaO is superior to sand as a bed material, using CaO-CLG can increase H_2 concentration from 48.07% obtained for Sand-CLG to 78% and reduce CO_2 concentration from 21.16% obtained for the same baseline system to almost zero (Fig. 6). The plausible reason of this CaO-based phenomenon could be explained by the correlation between the partial pressure reduction of CO_2 due to carbonation reaction (Eq. (2)) and the enhancement of water–gas shift reaction (Eq. (3)). More details of this are previously described in Section 3.2. In addition, the significant decrease of tar content consistent with the increase of total gas yield was also observed in CaO-CLG as compared to Sand-CLG (Fig. 7). This finding can be explained with catalytic tar reforming reaction of CaO (Eq. (4)). As observed in the present study, similar results of the influence of CaO on the production of H_2, CO_2 and tar were also reported by many authors [9], [29], [30] and [31] but those results were obtained from different gasification systems. Thus, to verify the superiority of CLG system, the following discussion comparatively presents the results obtained for CaO-CLG and CaO-BFBG, both of which were identically operated with CaO as a bed material.

Compared to CaO-BFBG, the CaO-CLG system is better with higher concentration and yields of H_2 and lower contents of CO_2 and tar (Fig. 6 and Fig. 7). This may be because of more availability of CaO in the gasifier of CaO-CLG. In CaO-CLG, the gasifier is continuously filled with regenerated CaO that is obtained from calcination of deactivated CaO (i.e., $CaCO_3$) in the regenerator, unlike CaO-BFBG in which deactivated CaO cannot be regenerated. This can contribute to the findings above.

CONCLUSIONS

This work investigated steam gasification of biomass with the presence of CaO in a uniquely designed CLG system for hydrogen production with in situ CO_2 capture and tar reduction. The effect of solid circulation rates on production of gas and tar was examined. Comparative study on gas and tar production from CaO-CLG, Sand-

CLG and CaO-BFBG was also conducted. Important conclusions drawn from the present study are as follows:

- By increasing the solid circulation rate from 0.91 to 1.14 kg/m^2s, the concentration and yield of H_2 obtained increased to 78% and 451.11 ml/g of biomass respectively at the circulation rate of 1.04 kg/m^2s. Also at this point, the lowest CO_2 concentration of 4.98% and the lowest tar content of 2.48 g/Nm^3 were obtained. Higher circulation rate of solid can enhance hydrogen production with simultaneous CO_2 and tar reduction due to more sites available for carbonation reaction and catalytic tar reforming reaction between intermediate products (CO_2 and tar) and regenerated CaO. However, too high circulation rate can cause inverse results due to a short residence time of $CaCO_3$ undergoing calcination reaction in the regenerator. This study suggests the solid circulation rate of 1.04 kg/m^2s to be optimal for the system.

- A comparative study of three different gasification systems shows that H_2 concentration and yield were highest in CaO-CLG, while the CO_2 concentration and tar content were lowest. Compared to Sand-CLG, CaO-CLG gave 30% higher concentration of H_2 and triple yield of H_2. Once compared to CaO-BFBG, it gave 15% higher concentration of H_2 and almost double yield of H_2. Besides, tar content obtained for CaO-CLG significantly reduced from 68.5 g/Nm^3 for Sand-CLG and 26.71 g/Nm^3 for CaO-BFBG to 2.48 g/Nm^3, while CO_2 concentration obtained for CaO-CLG also drastically reduced to nearly 0%. This study can indicate that the CaO-CLG system is a promising technology for hydrogen production with in situ CO_2 and tar reduction.

ACKNOWLEDGMENTS

The work was carried out in the Circulating Fluidized Bed Research laboratory of Dalhousie University, Canada in a Chemical Looping Gasification unit designed and built by Greenfield Research

Incorporated. The first author, Jakkapong Udomsirichakorn thanks the Energy Policy & Planning Office (EPPO) in Thailand for providing the King HRD scholarship for his PhD study at the Asian Institute of Technology and also acknowledges the financial support provided by PTT Public Company Limited (Thailand), as well as supports provided by Greenfield Research Incorporated and Dalhousie University for the experiments.

REFERENCES

1.   A. Tanksale, J.N. Beltramini, G.Q.M. Lu, A review of catalytic hydrogen production processes from biomass, Renewable and Sustainable Energy Reviews 14 (2010) 166–182.

2.   M. Asadullah, S. Ito, K. Kunimori, M. Yamada, K. Tomishige, Biomass gasification to hydrogen and syngas at low temperature: novel catalytic system using fluidizedbed reactor, Journal of Catalysis 208 (2002) 255–259.

3.   S. Chen, D. Wang, Z. Xue, X. Sun, W. Xiang, Calcium looping gasification for highconcentration hydrogen production with CO_2 capture in a novel compact fluidized bed: simulation and operation requirements, International Journal of Hydrogen Energy 36 (2011) 4887–4899.

4.   G.J. Stiegel, M. Ramezan, Hydrogen from coal gasification: an economical pathway to a sustainable energy future, International Journal of Coal Geology 65 (2006) 173–190.

5.   S. Yolcular, Hydrogen production for energy use in European Union countries and Turkey, Energy Sources, Part A: Recovery, Utilization, and Environmental Effects 31 (2009) 1329–1337.

6.   H. Balat, E. Kirtay, Hydrogen from biomass — present scenario and future prospects, International Journal of Hydrogen Energy 35 (2010) 7416–7426.

7.   Y. Kalinci, A. Hepbasli, I. Dincer, Biomass-based hydrogen production: a review and analysis, International Journal of Hydrogen Energy 34 (2009) 8799–8817.

8.   J. Guan, Q. Wang, X. Li, Z. Luo, K. Cen, Thermodynamic analysis of a biomass anaerobic gasification process for hydrogen production with sufficient CaO, Renewable Energy 32 (2007) 2502–2515.

9.   L. Han, Q. Wang, Y. Yang, C. Yu, M. Fang, Z. Luo, Hydrogen production via CaO sorption enhanced anaerobic gasification of sawdust in a bubbling fluidized bed, International Journal of Hydrogen Energy 36 (2011) 4820–4829.

10. J. Udomsirichakorn, P.A. Salam, Review of hydrogen-enriched gas production from steam gasification of biomass: the prospect of CaO-based chemical looping gasification, Renewable and Sustainable Energy Reviews 30 (2014) 565–579.

11. N.H. Florin, A.T. Harris, Enhanced hydrogen production from biomass with in situ carbon dioxide capture using calcium oxide sorbents, Chemical Engineering Science 63 (2008) 287–316.

12. N. Nipattummakul, I.I. Ahmed, A.K. Gupta, S. Kerdsuwan, Hydrogen and syngas yield from residual branches of oil palm tree using steam gasification, International Journal of Hydrogen Energy 36 (2011) 3835–3843.

13. M. Baratieri, P. Baggio, L. Fiori, M. Grigiante, Biomass as an energy source: thermodynamic constraints on the performance of the conversion process, Bioresource Technology 99 (2008) 7063–7073.

14. C. Franco, F. Pinto, I. Gulyurtlu, I. Cabrita, The study of reactions influencing the biomass steam gasification process, Fuel 82 (2003) 835–842.

15. S. Rapagná, H. Provendier, C. Petit, A. Kiennemann, P. Foscolo, Development of catalysts suitable for hydrogen or syn-gas production from biomass gasification, Biomass and Bioenergy 22 (2002) 377–388.

16. D. Ross, R. Noda, M. Horio, A. Kosminski, P. Ashman, P. Mullinger, Axial gas profiles in a bubbling fluidized bed biomass gasifier, Fuel 86 (2007) 1417–1429.

17. J. Udomsirichakorn, P. Basu, P.A. Salam, B. Acharya, Effect of CaO on tar reforming to hydrogen-enriched gas with in-process CO2 capture in a bubbling fluidized bed biomass steam gasifier, International Journal of Hydrogen Energy 38 (2013) 14495–14504.

18. M. Balat, M. Balat, E. Kirtay, H. Balat, Main routes for the thermo-conversion of biomass into fuels and chemicals, Part 2: gasification systems, Energy Conversion and Management 50 (2009) 3158–3168.

19. V. Manovic, D. Lu, E.J. Anthony, Steam hydration of sorbents from a dual fluidized bed CO2 looping cycle reactor, Fuel 87 (2008) 3344–3352.

20. L.S. Fan, Chemical Looping Systems for Fossil Energy Conversions, John Wiley and Sons, New York, 2010.

21. B. Acharya, A. Dutta, P. Basu, Chemical looping gasification of biomass for hydrogenenriched gas production with in-process carbon dioxide capture, Energy & Fuels 23 (2009) 5077–5083.

22. B. Acharya, A. Dutta, P. Basu, Circulating-fluidized-bed-based calcium-looping gasifier: experimental studies on the calcination–carbonation cycle, Industrial and Engineering Chemistry Research 51 (2012) 8652–8660.

23. B. Moghtaderi, Review of the recent chemical looping process developments for novel energy and fuel applications, Energy & Fuels 26 (2012) 15–40.

24. CEN/TS 15439:2006, Biomass Gasification — Tar and Particles in Product Gases — Sampling and Analysis, 2006.

25. P. Basu, Combustion and Gasification in Fluidized Beds, CRC/Taylor & Francis, Boca Raton, FL, 2006.

26. B. Acharya, A. Dutta, P. Basu, An investigation into steam gasification of biomass for hydrogen enriched gas production in presence of CaO, International Journal of Hydrogen Energy 35 (2010) 1582–1589.

27. P. Weerachanchai, M. Horio, C. Tangsathitkulchai, Effects

of gasifying conditions and bed materials on fluidized bed steam gasification of wood biomass, Bioresource Technology 100 (2009) 1419–1427.

28. C. Pfeifer, B. Puchner, H. Hofbauer, In-situ CO_2-absorption in a dual fluidized bed biomass steam gasifier to produce a hydrogen rich syngas, International Journal of Chemical Reactor Engineering 5 (2007) A9.

29. M.R. Mahishi, D.Y. Goswami, An experimental study of hydrogen production by gasification of biomass in the presence of a CO_2 sorbent, International Journal of Hydrogen Energy 32 (2007) 2803–2808.

30. S. Koppatz, C. Pfeifer, R. Rauch, H. Hofbauer, T. Marquard-Moellenstedt, M. Specht, H2 rich product gas by steam gasification of biomass with in situ CO_2 absorption in a dual fluidized bed system of 8 MW fuel input, Fuel Processing Technology 90 (2009) 914–921.

31. S. Rapagná, N. Jand, A. Kiennemann, P.U. Foscolo, Steam-gasification of biomass in a fluidized-bed of olivine particles, Biomass and Bioenergy 19 (2000) 187–197.

Chapter 3

Bubbling Fluidised Bed Gasification of Wheat Straw–Gasifier Performance Using Mullite as Bed Material

Seán T. Mac an Bhaird[a], Phil Hemmingway[a], Eilín Walsh[a], Amado L. Maglinao[b], Sergio C. Capareda[b], and Kevin P. McDonnell[c]

[a]School of Biosystems Engineering, University College Dublin, Belfield, Dublin 4, Ireland

[b]Department of Biological and Agricultural Engineering, Texas A&M University, College Station, TX, USA

[c]School of Agriculture and Food Science, University College Dublin, Belfield, Dublin 4, Ireland

ABSTRACT

The adoption of wheat straw as a fuel for gasification processes has been hindered due to a lack of experience and its propensity to cause bed agglomeration in fluidised bed gasifiers. In this study wheat straw was gasified in a small scale, air blown bubbling fluidised bed using mullite as bed material. The gasifier was successfully operated and isothermal bed conditions maintained at temperatures up to 750 °C. Below this temperature, the gasifier was operated at equivalence ratios from 0.1 to 0.26. The maximum lower heating value of the producer gas was approximately 3.6 MJ m^{-3} at standard temperature and pressure (STP) conditions and was obtained at an equivalence ratio of 0.165. In general, a producer gas with a lower heating value of approximately 3 MJ m^{-3} at STP could be obtained across the entire range of equivalence ratios operated. The lower heating value tended to fluctuate, however, and it was considered more appropriate for use in heat applications than as a fuel for internal combustion engines. The concentration of combustibles in the producer gas was lower than that obtained from the gasification of wheat straw in a dual distributor type gasifier and a circulating fluidised bed. These differences were associated with reactor design and, in the case of the circulating fluidised bed, with higher temperatures. Equilibrium modelling at adiabatic conditions, which provides the maximum performance of the system, showed that the gasifier was operating at suboptimal equivalence ratios to achieve greatest efficiencies. The maximum calculated theoretical cold gas efficiency of 73% was obtained at an equivalence ratio of 0.35.

INTRODUCTION

The Irish farming sector produces significant surpluses of cereal straw each year and the potential exists to use this biomass in energy production: the feasible straw resource for energy generation has been estimated at 80,000–320,000 tonnes annually (Henriksen et al., 2006, Stahl et al., 2004 and Yin, 2011). There are obstacles to this use, however, amongst which is the lack of experience in this area (Hongli et al., 2009). This study investigates the use of cereal straw, specifically wheat straw, as a fuel for gasification. Gasification is a thermochemical process that has the potential to convert solid fuel into a gas with various end uses. The quality of the gas determines its most appropriate end use which spans both energy applications and chemical synthesis. The firing of the gas in boilers or its use in heat applications has been identified as the simplest end use (Maniatis, 2001).

Air, oxygen, steam, carbon dioxide, and hydrogen can all be used as gasification agents (Ren et al., 2010 and Vigouroux, 2001). When using air or oxygen as the gasification agent, a portion of the feedstock is combusted to drive the endothermic gasification reactions. The portion of feedstock that undergoes complete combustion is dependent on the equivalence ratio (ER) which is the ratio of the actual air-fuel ratio to that required for complete combustion. The ER is a key parameter affecting the heating value of gas produced by gasification, and in many cases is the dominant operating parameter (de Jong et al., 2003 and van der Drift et al., 2001). Specific gas yields are reported to increase continuously with ER (Mansaray et al., 1999, Natarajan et al., 1998, Sheth and Babu, 2009 and Zainal et al., 2002). Generally, at low ERs the heating value of the gas increases with ER before peaking and then starting to decrease (Mansaray et al., 1999, Sheth and Babu, 2009 and Zainal et al., 2002). This decrease is associated with a reduction in combustible gases and dilution with carbon dioxide and nitrogen.

Straw gasification has been shown to cause operational difficulties, for example the low bulk density of straw makes it unsuitable for gasification in moving bed gasifiers (Maniatis, 2001). Fluidised bed gasifiers can accommodate straw which has not first been densified, however its gasification in fluidised beds has been reported to cause severe ash sintering and bed agglomeration (Maniatis, 2001). Gasification temperature has been identified as the most important parameter in relation to the propensity of a bed to agglomerate, with higher temperatures increasing the risk (Natarajan et al., 1998). In autothermal gasification, where the heat required by the net endothermic gasification reactions is delivered by combusting a portion of the fuel, the temperature is controlled by the ER: the greater the ER, the greater the temperature. Ideal and theoretical gasification has been identified to occur at ERs between

0.19 and 0.43 (Zainal et al., 2002), thus by limiting the ER to avoid bed agglomeration the performance of the gasifier may be affected.

This study investigates the gasification of wheat straw in a small scale, air blown bubbling fluidised bed (BFB) using mullite, an alumina sand that has been reported to increase agglomeration temperatures (Ergudenler and Ghaly, 1993 and van der Drift and Olsen, 1999), as bed material. The most appropriate use of the gas produced will then be determined based on the higher heating value.

MATERIALS AND METHODS

Apparatus and Operating Conditions

A bubbling fluidised bed gasifier developed by Texas A&M University (Parnell Jr. and LePori, 1988) was used to study the gasification characteristics of wheat straw. The gasifier is a 1-foot diameter, skid-mounted fluidised bed gasifier with a feed rate rating of approximately 1.8 tonnes day^{-1} (150 lb h^{-1}) (Capareda and Maglinao, 2009). A schematic depiction of the gasifier is provided in Fig. 1 (after Capareda and Maglinao, 2009).

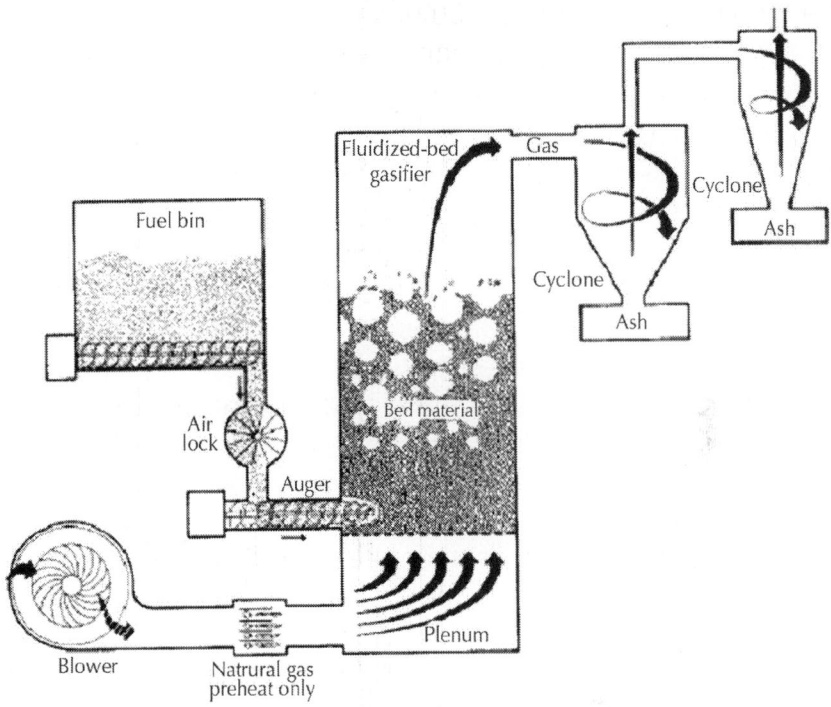

Figure 1: Schematic of fluidised bed gasifier (after Capareda and Maglinao, 2009).

The gasifier was designed to accommodate cotton gin trash and similar biomass without the need for pre-treatment. A range of feedstocks have been successfully gasified to date using the gasifier including poultry litter, wood chips, cotton gin trash, dairy manure, sorghum, and switchgrass (Capareda and Maglinao, 2009, Maglinao, 2009, Maglinao and Capareda, 2008 and Parnell Jr. and LePori, 1988). The gasification agent utilised in all studies was air. The design power output of the gasifier was set at approximately 73 kW based on the heating value of the producer gas from the gasification of cotton gin trash.

A natural gas burner is employed to preheat the air to the gasifier and achieve operating temperatures. The fuel is fed into the gasifier via an auger system; an optical counter is used to measure the rotational speed of the screw conveyor system feeding the fuel directly into the bottom of the bed close to the distribution plate. This feed configuration minimises segregation of the fuel from the bed which has a detrimental effect on the performance of a gasifier: segregation is linked to higher levels of tar, reduced carbon conversion efficiency, and potentially increased agglomeration (Narváez et al., 1996 and Salour et al., 1993). Fuel segregation is known to occur with straw and it is recommended that the fuel be fed directly into the bed (Salour et al., 1993).

The producer gas exiting the gasifier passes through two cyclones in series: the first cyclone is designed to reduce the particulate loading to a maximum of 3 g m^{-3} and the second to a maximum 0.5 g m^{-3}. The maximum pressure drop across the two cyclones is 203 mm of water. Collection bins for the cut particles are located at the bottom of each cyclone. The turbulent, fluidised state of inert particles in the bed creates a near isothermal zone and enables accurate control of reaction temperatures (Capareda and Maglinao, 2009). To record reaction temperatures, a K-type thermocouple (Omega CAIN-14U: Omega Engineering Inc., Stamford, CT, USA) is located just below the bed base with three further thermocouples placed along the height of the bed at 152.4, 254, and 469.4 mm above the bed base. Readings from these thermocouples are referenced as T1–T4, respectively. Pressure readings are taken at the base and the upper base of the fluidised bed by differential pressure transmitters (Omega PX274: Omega Engineering Inc., Stamford, CT, USA and Dwyer Series 677: Northeast Controls Inc., Upper Saddle River, NJ, USA). Pressure and temperatures are continuously monitored and logged using a data logger (Omega OM-320: Omega Engineering Inc., Stamford, CT, USA).

A laminar flow element is located between the blower and the gasifier to determine the primary air flow rate to the gasifier. An online gas analyser (HORIBA Scientific, Irvine, CA, USA) located after the second cyclone was used to provide dynamic measurements of the composition of the producer gas in terms of total methane, carbon monoxide, carbon dioxide, and hydrogen. Gas samples were collected after the second cyclone in 1 litre Tedlar bags for gas chromatography analysis (Model 310: SRI Instruments, Torrance,

CA, USA). For gas collection, a small diameter pipe system which extended horizontally from the main gas flow was used to channel a proportion of the gas to the online analyser/Tedlar bag. The lower heating value (LHV) of the producer gas was calculated from the gas compositional readings.

Equilibrium modelling was conducted to determine the maximum theoretical efficiency of the system (Channiwala and Parikh, 2002, Mahishi and Goswami, 2007 and Prins et al., 2007). Equilibrium modelling results were compiled using HSC Chemistry© 6.1 chemical reaction and equilibrium software (Outotec Research Oy, Pori, Finland). The modelling procedure is based on the principle of Gibbs free energy minimisation to estimate the expected chemical composition of the producer gas yield from user defined species. The gasification model adopted the fuel characteristics and operating conditions of the gasifier and accounted for the following output species: solid carbon (C), carbon monoxide (CO), carbon dioxide (CO_2), hydrogen (H_2), water (H_2O), oxygen (O_2), nitrogen (N_2), argon (Ar), methane (CH_4), ethylene (C_2H_4), and ethane (C_2H_6). The cold gas efficiency (CGE) of the gasifier during each of the tests was calculated according to the energy content of the producer gas and the energy content of the wheat straw fuel, as described by van der Drift et al. (2001).

Feedstock and Bed Material

Wheat straw sourced in Texas, United States of America, was utilised for all experiments. Proximate and ultimate analyses of the fuel were conducted in accordance with Maglinao and Capareda (2008). To allow for ready fuel feeding into the gasifier, a hammer mill was used to reduce the size of the fuel to enhance feeding. Resource constraints did not allow for size characterisation of the wheat straw fuel to be conducted, however LePori and Soltes (1985) reported that variations in composition and particle size can be overcome by the violent agitation of solids in a fluidised bed gasifier, which provides efficient conversion reactions.

Mullite, an alumina sand that has been used to counteract agglomeration tendencies in fluidised beds (Ergudenler and Ghaly, 1993), was used as bed material. Its adoption has seen the successful gasification of wheat straw in a BFB gasifier of novel dual distributor type design at temperatures greater than 900 °C (Ergudenler and Ghaly, 1993). The mullite was passed through a 425 μm USA standard sieve meeting ASTM E 11 specification and the gasifier was filled to just below the disengagement zone, equal to 40 kg of bed material added to the gasifier. The purity and composition of the mullite are shown in Table 1 and the particle size distribution of the mullite is shown in Table 2. Fresh bed material was used for each gasification test. In total, three gasification tests were conducted; the average bed temperatures and ERs are detailed inTable 3.

Table 1: Purity and composition of mullite used as bed base (after C-E Minerals, 2013)

Mulcoa 47	
Al_2O_3	46.8 (min 46.0)
SiO_2	50.0
TiO_2	1.89
Fe_2O_3	0.95 (max 1.0)
CaO	0.04
MgO	0.08
Na_2O	0.09
K_2O	0.09
P_2O_5	0.09
Mineralogy	
%Mullite	65
%Glass	20
%Cristabolite	15

Table 2: Particle size distribution of mullite used as bed material

USS SieveGrade	8 2.36 mm	121.70 mm	14 1.40 mm	20850 μm	30600 μm	40425 μm	50300 μm	70212 μm	100150 μm	140106 μm	20075 μm	27053 μm	32545 μm	PAN[a]
10 × 18[1]	0–3	10–25	45–65	60–82	0–15									0–5
10 × 28[1]	0	15–25	45–65	6–16	5–15	0–6								0–2
14 × 28[1]		0	1 max	30–55	35–45	10–25	5 max							1.5 max
16 × 30[1,2]			0–3	65–75		4 max								1 max
22S[1,2]			TR	15–25	32–47	27–37	4–10						3 max	TR
35S[1,2]			TR	1–5	21–38	40–54		9–19		2–8				3 max
50S[1,2]				0	1–9	22–37	26–40	12–22	6–16	1–6				3 max
60S[1,2]					0	0–5	30–48	30–44	9–22	2–7				3 max
20 × 50[1,2]			TR	0–8	20–50		50–72						2 max	TR
25 × 80[1,2]				0–5	20 min 30 avg	80–93	7–12						3 max	
50 × 100[1]						TR	5–20		70–86	0–15	3 max		1 max	TR
60 × 200[1]							0	0–11	65–90		5–20	0–6		3 max
200 IC-C[1,2,3]									TR		15–25			75–85
325 IC-C[1,2,3]									TR				5–15	85–95

Table 3: Average operating conditions adopted in the three gasification tests investigating the gasification of wheat straw in a small scale, air blown bubbling fluidised bed gasifier

Test	Fuel	Average equivalence ratio	Average bed temperature (°C)[a]
1	Wheat straw	0.27	770
2	Wheat straw	0.23	728
3	Wheat straw	0.17	654

[a]The bed temperature is taken as the average of the readings from thermocouples T2, T3, and T4.

Procedures

Gasification commenced by fluidising the mullite bed by gradually increasing the volume flow rate of air entering the gasifier until a decrease in total pressure drop followed by its characteristic flattening out was evident. Fluidisation was achieved at superficial gas velocity of 0.27 m s^{-1} and was maintained throughout the experiment. At this point, the bed was fully fluidised and bubbling vigorously. The air was preheated with a natural gas burner to increase the reactor temperature to 500 °C. After reaching 500 °C, the natural gas burner was turned off and fuel (wheat straw) was introduced into the reactor. Fuel was then used to further heat up the reactor until the desired operating temperature was reached.

After fuel feeding had commenced and the gas burner had been turned off, the fuel feed rate was steadily decreased to obtain the desired operating conditions of temperature and equivalence ratio. The desired air-to-fuel ratio was obtained by adjusting the speed of the screw conveyor auger of the feeding system and the air flow (Maglinao and Capareda, 2010). For the first two tests the target temperature was 700 °C. For the third test the target temperature was lower to avoid loss of isothermal conditions. In this test, the fuel feed was initiated when the target temperature of 620 °C was reached and the fuel was fed at an initial ER of approximately 0.35 before being reduced to 0.15 over a short period of time.

The design of the gasifier did not allow for ash to be removed directly from the gasifier chamber for inventory control. During operation, the level of char in the collection bins located at the bottom of each cyclone was continuously monitored and emptied when required. The char was transferred to stainless steel containers which were sealed to reduce air exposure and the potential oxidation of the hot char.

The composition of the gas was measured in Tests 1 and 3 by the online gas analyser. The online gas analyser was calibrated utilising standard gases and ambient air before each test. For offline gas analysis conducted in Test 2, the gas chromatograph was calibrated utilising standard gases. Once a gas sample had been collected, the composition analysis was conducted in triplicate ($n = 3$) as soon as possible thereafter to minimise any gas egress. The gas composition was then compared to that from the online gas analyser used in Tests 1 and 3 to further ensure the accuracy of the calibration of the online gas analyser. Following each gasification test, and once the gasifier had cooled, the interior of the reactor was inspected and the bed material removed via vacuum; any material that had adhered to the walls of the reactor was removed with a wire brush. The reactor was subsequently filled with fresh bed material for the next test.

Heating Value and End Use of the Gas

The composition of the gas from each test was used to determine the higher heating value of the produced gas, and from this the lower heating value was calculated. The most appropriate use of the gas (i.e. in energy applications or chemical synthesis) was then determined.

RESULTS AND DISCUSSION

Gasifier Operation

Fig. 2 and Fig. 3 show the temperature and pressure readings recorded during the three gasification tests.

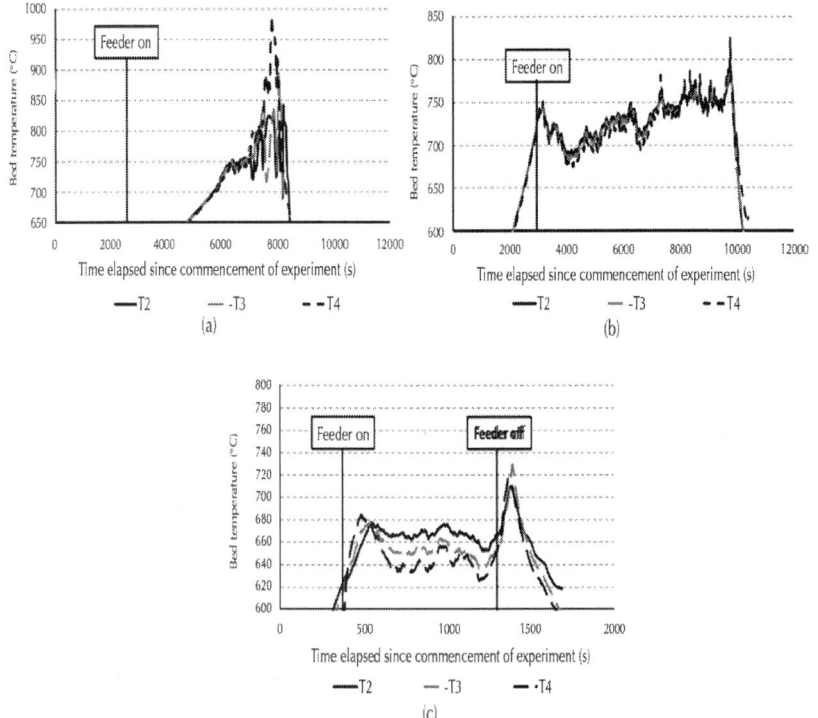

Figure 2: Temperature readings within the mullite bed of a small scale, air blown bubbling fluidised bed gasifier when gasifying wheat straw during (a) Test 1; (b) Test 2; and (c) Test 3. Temperature readings T2, T3, and T4 from thermocouples positioned 152.4 mm, 254 mm, and 469.4 mm above the bed base, respectively.

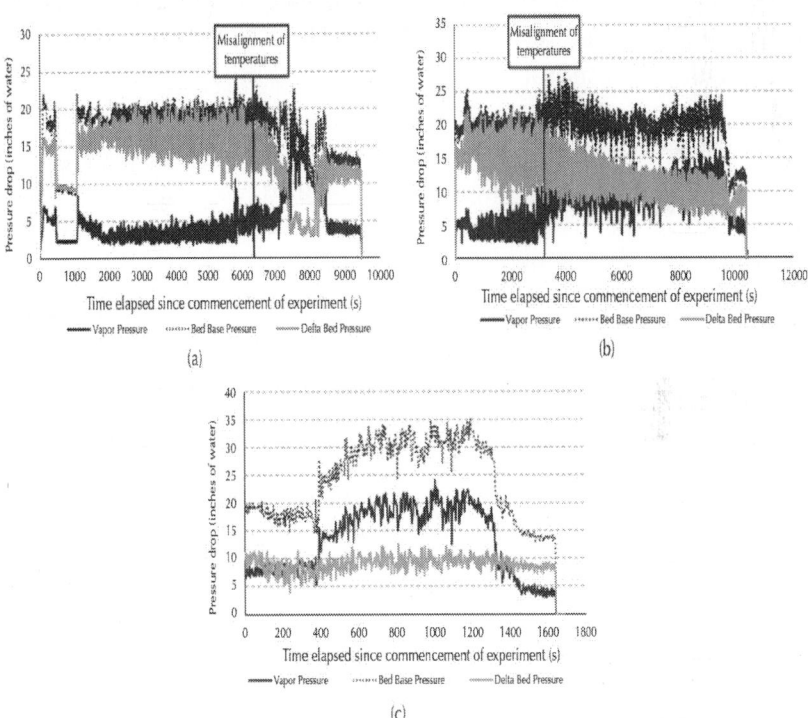

Figure 3: Pressure readings from within the air blown bubbling fluidised bed gasifier during gasification of wheat straw during (a) Test 1; (b) Test 2; and (c) Test 3.

As stated earlier, higher gasification temperatures were achieved in Tests 1 and 2 than in Test 3. In Tests 1 and 2, the dynamics of the bed were observed to be affected when temperatures rose above 750 °C: at approximately 750 °C, the temperatures along the height of the bed were seen to align before the temperature of the upper bed rose and the highest temperature was recorded there (note the significant change in temperature recorded by T4 thermocouple in Fig. 2a), indicating stratified combustion of biomass above the bed and poor bed fluidisation. This was followed by a pronounced decrease in the pressure differential across the bed, which occurred 2299 s after the upper bed became the hottest region of the bed in Test 1. In Test 2, a similar pronounced pressure drop was evident much earlier, at 830 s after the temperature in the upper bed rose, though not as substantially as in Test 1 (Fig. 2b). These temperature and pressure excursions suggest channelling within the bed, allowing fuel to pass through the bed and combust in the vapour space above it (Scala and Chirone, 2008). The data indicate that changes in temperature alignments within the bed point to the onset of defluidisation at approximately 750 °C, beyond which a detrimental effect on the dynamics of the bed was evident; this temperature closely corresponds to the 760 °C reported by Salour et al. (1993) during gasification of wood and rice straw.

For Test 3, during which the gasifier was operated at a lower temperature of 620 °C, the temperatures within the bed did not exceed 683 °C and reduced with height above the bed. The temperatures along the height of the bed were seen to track one another: the average difference in temperature readings from T2 and T4 remained less than 3.5%. Such isothermal conditions are a known advantage of fluidised bed gasifiers over fixed bed designs (Bridgwater, 2003) and demonstrate their superior heat and material transfer characteristics between gaseous and solid phases (Warnecke, 2000). In Test 3, no large sustained or sudden variations in pressures within the gasifier were evident. These temperature and pressure readings indicate stable gasification conditions.

The gasifier was operated at higher ERs in Tests 1 and 2 (0.27 and 0.23, respectively) than in Test 3 (0.17). The average ER of Test 3 includes the initial heating period where fuel was fed at a relatively high ER for 492 s in order to achieve the desired operating temperature. When this initial heating period is omitted the average ER was 0.15. This is a relatively low ER for biomass gasification; ideal and theoretical gasification has been identified to occur at ERs of between 0.19 and 0.43 (Zainal et al., 2002), and is considerably lower than the ER of 0.25 associated with maximising the mole fraction of combustibles in air gasification of wheat straw (Ergudenler and Ghaly, 1993). Up to the point of temperature convergence observed in Tests 1 and 2, the gasifier was operated at an average ER of 0.20 and 0.22, respectively.

Cold Gas Efficiency

Equilibrium models represent the maximum gasification efficiency (Channiwala and Parikh, 2002 and Mahishi and Goswami, 2007), however it is deemed difficult to match equilibrium performance for gasification temperatures lower than 1000 °C (Prins et al., 2007). It was seen that under equilibrium conditions there is a general increase in cold gas efficiency (CGE) with ER (a trend also reported in literature). The maximum CGE of 73% was obtained at an ER of 0.35. This point of operation coincides with the carbon boundary point, the point at which carbon is no longer produced by the system. This maximum CGE occurs at a higher ER than that at which the gasifier was operated: 0.27, 0.21, and 0.17 for Tests 1–3, respectively (Fig. 4). This indicates that the operating conditions were sub-optimal, however were deemed necessary to ensure isothermal conditions were maintained during gasification.

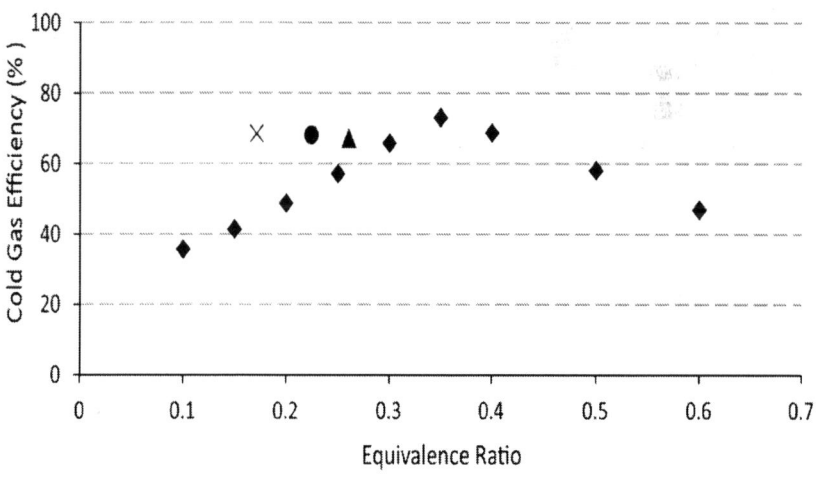

Figure 4: Cold gas efficiency of modelled gasification vs. the ER of gasification tests.

Heating Value and End Use of the Gas

The proximate and ultimate composition of the wheat straw gasified in the BFB gasifier is reported in Table 4.

Table 4: Proximate and ultimate composition of the wheat straw gasified in this study

Fuel	Wheat straw
Moisture content (wt%) as received	8.02 ± 0.28
Volatile content (wt%) as received	81.03
Fixed carbon (FC) (%)	8.21
C content (%)	43.51
H content (%)	5.43
N content (%)	0.58
O content (%)	39.13
S content (%)	0.12
Ash content (%)	11.24

Feedstock moisture content is known to have a substantial effect on the efficiency of gasification which can be either positive or negative depending on the moisture content of the feedstock. Asadullah (2014) reported that under the temperatures achieved during gasification, the water contained in the feedstock is converted to steam, a gasification agent used to convert volatiles to producer gas. If the moisture content is excessively high, however, the efficiency of the process is reduced as the moisture absorbs a greater proportion of heat, increasing the energy requirement to convert the biomass into producer gas (Asadullah, 2014). In this case, the moisture content of the straw as received was 8.02 ± 0.28%, which is considered acceptable for gasification as Kaewluan and Pipatmanomai (2011) successfully co-gasified feedstocks with moisture contents as high as 27% in a fluidised bed gasifier.

Fig. 5 illustrates the HHV of the gas produced during this experimental work and compares it with the HHV of the gas produced during the gasification of cotton gin trash using the same BFB gasifier (Craig, 1980). It is evident that the HHV of the gas produced from cotton gin trash is generally greater; at lower fuel-to-air ratios a convergence with the HHVs recorded in this experimental work occurs, however. The HHVs from this work are also noted to be lower than the 5.59 MJ m⁻³ reported by Capareda and Maglinao (2009) when the BFB gasifier was fuelled with poultry litter and wood chips at a similar temperature (760 °C vs. 770 °C in Test 1) as well as the HHVs associated with the gasification of agricultural residues in BFBs at a similar temperature (4.85–5.87 MJ m⁻³ at 760 °C; Salour et al., 1993).

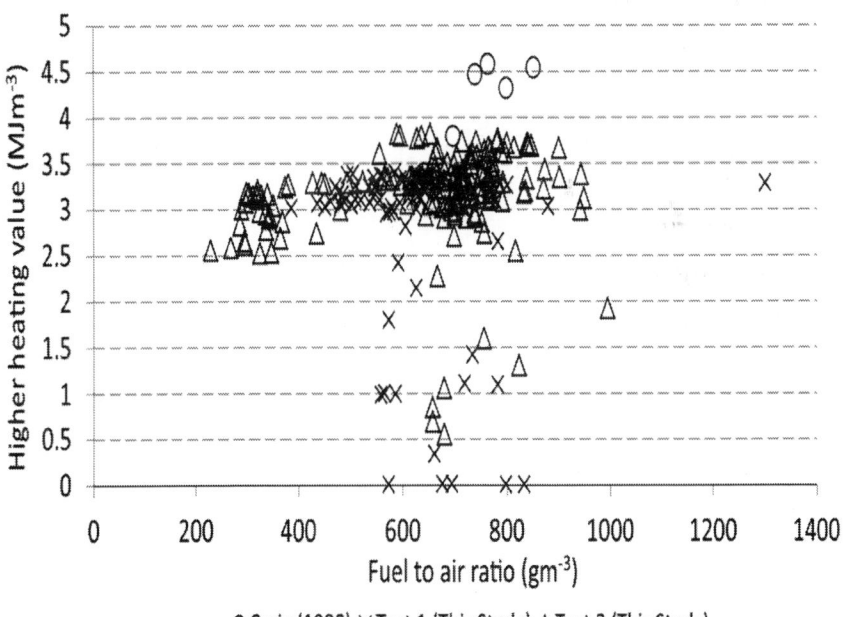

O Craig (1980) X Test 1 (This Study) △ Test 3 (This Study)

Figure 5: A comparison of the higher heating values of the gas obtained from the gasification of wheat straw and cotton gin trash in an air blown bubbling fluidised bed gasifier.

The gasification of wheat straw in the BFB gasifier in this study produced a maximum LHV of 3.6 MJ m^{-3} at an ER of 0.165. A producer gas with LHV of approximately 3 MJ m^{-3} at standard conditions of temperature and pressure (STP) was obtainable at a range of ERs. In all tests, the LHV of the gas exceeded the minimum heating value required for use in internal combustion engines, reported by Knoef (2005) to be 2.5 MJ m^{-3} at STP. Knoef (2005) stated that a heating value in excess of 4.2 MJ m^{-3} at STP is preferable; indeed, literature indicates that engines are typically run on gases meeting this more demanding limit. Although the LHV at STP justifies the use of the produced gas as a fuel for internal combustion engines, the necessary cooling of the gas to STP for use in internal combustion engines (Bridgwater, 1995 and Stevens, 2001) can cause the condensation of tars which can lead to engine failure (Stevens, 2001). An average tar value of 10 g m^{-3} at STP has been reported for fluidised bed gasifiers (Milne et al., 1998). To facilitate the use of the gas obtained in these tests in internal combustion engines, secondary tar reduction measures would therefore be required to reduce the tar loading to the tolerance level of internal combustion engines, reported to range from as low as 1 mg m^{-3} at STP to 100 mg m^{-3} at STP (Bhattacharya et al., 2001, Boerrigter and Rauch, 2006, Cao et al., 2006, Han and Kim, 2008 and Lettner et al., 2007).

In contrast to the use in internal combustion engines of gas obtained from the gasification of wheat straw, only particulates must be reduced for use in heat applications (Maniatis, 2001). Such applications have been identified as the simplest use of producer gas, nonetheless there are very few examples (Maniatis, 2001). The Kymijärvi power plant in Finland, operated by Lahden Lämpövoima Oy, is an example of a gasification project where low calorific producer gas (typically between 1.6 and 2.4 MJ m^{-3}) has been successfully co-fired in a coal-fired boiler using a fluidised bed of circulating design to produce electricity for the owner and district heat for the city of Lahti (Raskin et al., 2001). The LHV of the gas produced in this study exceeds this range, thus heat applications are considered to be the most suitable end use for this gas rather than its use in internal combustion engines.

CONCLUSIONS

In order to maintain isothermal bed conditions, the temperature of the bed was maintained below 750 °C; under these temperature conditions the gasifier was operated at ERs from 0.1 to 0.27. Equilibrium modelling showed that the gasifier was operating at suboptimal ERs: the maximum calculated theoretical CGE of 73% was obtained at an ER of 0.35.

The maximum LHV of the producer gas was approximately 3.6 MJ m^{-3} and was obtained at an ER of 0.165, well below the modelled ER to obtain maximum CGE. It was observed that, despite fluctuations, a producer gas with an LHV of approximately 3 MJ m^{-3} at STP could be obtained across the range of ERs. Considering the LHV obtained, and taking fluctuations in LHV into account as well as the reported tar content for fluidised bed gasifiers, it is suggested that the optimal use for the producer gas is in heat applications.

ACKNOWLEDGMENTS

This study was funded under the Charles Parsons Energy Research Award of Science Foundation Ireland[Grant Number 6C/CP/E001] supported by the Department of Communications, Energy and Natural Resources of the Government of Ireland.

REFERENCES

1.   Asadullah, M., 2014. Barriers of commercial power generation using biomass gasification gas: a review. Renew. Sustain. Energy Rev. 29, 201–215.

2.   Bhattacharya, S.C., Shwe Hla, S., Pham, H.-L., 2001. A study on a multi-stage hybrid gasifier-engine system. Biomass Bioenergy 21, 445–460.

3. Boerrigter, H., Rauch, R., 2006. Review of Applications of Gases from Biomass Gasification. ECN Energy Research Foundation, Petten, the Netherlands.

4. Bridgwater, A.V., 1995. The technical and economic feasibility of biomass gasification for power generation. Fuel 74, 631–653.

5. Bridgwater, A.V., 2003. Renewable fuels and chemicals by thermal processing of biomass. Chem. Eng. J. 91, 87–102.

6. C-E Minerals, 2013. MULGRAIN® Grains and Flours. Roswell, Georgia. Cao, Y., Wang, Y., Riley, J.T., Pan, W.-P., 2006. A novel biomass air gasification process for producing tar-free higher heating value fuel gas. Fuel Process. Technol. 87, 343–353.

7. Capareda, S., Maglinao, A., 2009. Animal manure and other biomass residue conversion into useful energy via fluidized bed gasification. In: Texas Animal Manure Management Issues Conference, Round Rock, TX.

8. Channiwala, S.A., Parikh, P.P., 2002. A unified correlation for estimating HHV of solid, liquid and gaseous fuels. Fuel 81, 1051–1063.

9. Craig, J.D., (MS Thesis) 1980. Performance of Gas Cleanup Criterion for a Cotton-Gin Waste Fluidized-Bed Gasifier. Texas A&M University, Texas, USA.

10. De Jong, W., Ünal, Ö., Andries, J., Hein, K.R.G., Spliethoff, H., 2003. Biomass and fossil fuel conversion by pressurised fluidised bed gasification using hot gas ceramic filters as gas cleaning. Biomass Bioenergy 25, 59–83.

11. Ergudenler, A., Ghaly, A.E., 1993. Agglomeration of alumina sand in a fluidized bed straw gasifier at elevated temperatures. Bioresour. Technol. 43, 259–268.

12. Han, J., Kim, H., 2008. The reduction and control technology of tar during biomass gasification/pyrolysis: an overview. Renew. Sustain. Energy Rev. 12, 397–416.

13. Henriksen, U., Ahrenfeldt, J., Jensen, T.K., Gøbel, B., Bentzen, J.D., Hindsgaul, C., Sørensen, L.H., 2006. The design,

construction and operation of a 75 kW two-stage gasifier. Energy 31, 1542–1553.

14. Hongli, W., Yitai, M., Minxia, L., 2009. Research on high efficient straw gasifier. In: Goswami, D.Y., Zhao, Y. (Eds.), Proceedings of ISES World Congress 2007, Vols. I–V. Springer, Berlin/Heidelberg, pp. 2383–2387.

15. Kaewluan, S., Pipatmanomai, S., 2011. Gasification of high moisture rubber woodchip with rubber waste in a bubbling fluidized bed. Fuel Process. Technol. 92, 671–677.

16. Knoef, H., 2005. Gaseous emissions and emission regulations. In: IEA Bioenergy International Workshop on Health, Safety and Environment of Biomass Gasification, 28th September, Innsbruck, Austria. IEA Bioenergy, pp. 20–29.

17. LePori, W.A., Soltes, E.J., 1985. Thermochemical conversion for energy and fuel. In: Hiler, E.A., Stout, B.A. (Eds.), Biomass Energy: A Monograph. Texas A&M University Press, College Station, TX, pp. 10–49.

18. Lettner, F., Timmerer, H., Haselbacher, P., 2007. Deliverable 8: Biomass Gasification—State of the Art Description. Intelligent Energy - Europe, Graz, Austria.

19. Maglinao, A.L., (MS Thesis) 2009. Instrumentation and Evaluation of a Pilot Scale Fluidized Bed Biomass Gasification System. Texas A&M University, Texas.

20. Maglinao, A.L., Capareda, S.C., 2010. Development of computer control system for the pilot scale fluidized bed biomass gasification system. In: ASABE Annual International Meeting, June 20–23, Pittsburgh, PA.

21. Maglinao, A.L., Capareda, S.C., 2008. Operation of the TAMU fluidized bed gasifier using different biomass feedstock. In: ASABE Annual International Meeting, June 29–July 2, Providence, Rhode Island.

22. Mahishi, M.R., Goswami, D.Y., 2007. Thermodynamic optimization of biomass gasifier for hydrogen production. Int. J. Hydrogen Energy 32, 3831–3840.

23. Maniatis, K., 2001. Progress in biomass gasification: an overview. In: Bridgwater, A.V. (Ed.), Progress in Thermochemical Biomass Conversion, Vol. 1. Blackwell Science Ltd, Oxford, UK, pp. 1–31.

24. Mansaray, K.G., Ghaly, A.E., Al-Taweel, A.M., Hamdullahpur, F., Ugursal, V.I., 1999. Air gasification of rice husk in a dual distributor type fluidized bed gasifier. Biomass Bioenergy 17, 315–332.

25. Milne, T.A., Evans, R.J., Abatzoglou, N., 1998. Biomass Gasifier "Tars": Their Nature, Formation, and Conversion. National Renewable Energy Laboratory, Golden, CO.

26. Narváez, I., Orío, A., Aznar, M.P., Corella, J., 1996. Biomass gasification with air in an atmospheric bubbling fluidized bed, Effect of six operational variables on the quality of the produced raw gas. Ind. Eng. Chem. Res. 35, 2110– 2120.

27. Natarajan, E., Nordin, A., Rao, A.N., 1998. Overview of combustion and gasification of rice husk in fluidized bed reactors. Biomass Bioenergy 14, 533–546.

28. Parnell Jr., C.B., LePori, W.A., 1988. System and process for conversion of biomass into usable energy. US Patent No. 4,848,249.

29. Prins, M.J., Ptasinski, K.J., Janssen, F.J.J.G., 2007. From coal to biomass gasification: comparison of thermodynamic efficiency. Energy 32, 1248–1259.

30. Raskin, N., Palonen, J., Nieminen, J., 2001. Power boiler fuel augmentation with a biomass fired atmospheric circulating fluid-bed gasifier. Biomass Bioenergy 20, 471–481.

31. Ren, Q., Zhao, C., Wu, X., Liang, C., Chen, X., Shen, J., Wang, Z., 2010. Formation of NOx precursors during wheat straw pyrolysis and gasification with O2 and CO2. Fuel 89, 1064–1069.

32. Salour, D., Jenkins, B.M., Vafaei, M., Kayhanian, M., 1993. Control of in-bed agglomeration by fuel blending in a pilot scale straw and wood fueled AFBC. Biomass Bioenergy 4, 117–133.

33. Scala, F., Chirone, R., 2008. An SEM/EDX study of bed agglomerates formed during fluidized bed combustion of three biomass fuels. Biomass and Bioenergy 32 (3), 252–266.
34. Sheth, P.N., Babu, B.V., 2009. Experimental studies on producer gas generation from wood waste in a downdraft biomass gasifier. Bioresour. Technol. 100, 3127–3133.
35. Stahl, R., Henrich, E., Gehrmann, H.J., Vodegel, S., Koch, M., 2004. Definition of a Standard Biomass. Forschungszentzrum Karlsruhe GmbH, Karlsruhe, Germany.
36. Stevens, D.J., 2001. Hot Gas Conditioning: Recent Progress with Larger-Scale Biomass Gasification Systems. Update and Summary of Recent Progress. National Renewable Energy Laboratory, Golden, Colorado.
37. Van der Drift, A., Olsen, A., 1999. Conversion of Biomass, {Private} Prediction and Solution Methods for Ash Agglomeration and Related Problems. ECN Energy Research Foundation, Petten, the Netherlands.
38. Van der Drift, A., van Doorn, J., Vermeulen, J.W., 2001. Ten residual biomass fuels for circulating fluidized-bed gasification. Biomass Bioenergy 20, 45–56.
39. Vigouroux, R.Z., 2001. Pyrolysis of Biomass: Rapid Pyrolysis at High Temperatures, Slow Pyrolysis for Active Carbon Preparation. Department of Chemical Engineering and Technology, Royal Institute of Technology, Stockholm, Sweden.
40. Warnecke, R., 2000. Gasification of biomass: comparison of fixed bed and fluidized bed gasifier. Biomass Bioenergy 18, 489–497.
41. Yin, C.-Y., 2011. Prediction of higher heating values of biomass from proximate and ultimate analyses. Fuel 90, 1128–1132.
42. Zainal, Z.A., Rifau, A., Quadir, G.A., Seetharamu, K.N., 2002. Experimental investigation of a downdraft biomass gasifier. Biomass Bioenergy 23, 283–289.

Chapter **4**

Assessment of Chemical Looping-Based Conceptual Designs for High Efficient Hydrogen and Power Co-generation Applied to Gasification Processes

Calin-Cristian Cormos, Ana-Maria Cormos, and
Letitia Petrescu

Babes-Bolyai University, Faculty of Chemistry and Chemical
Engineering, 11 Arany Janos Street, RO-400028, Cluj-Napoca,
Romania

ABSTRACT

Carbon capture and storage (CCS) technologies are expected to play a significant role in the coming decades for curbing the greenhouse gas emissions and to ensure a sustainable development of power generation and other energy-intensive industrial sectors. Chemical looping systems are very promising options for intrinsically capture CO_2 with lower cost and energy penalties. Gasification offers significant advantages compared with other technologies in term of lower energy and cost penalties for carbon capture, utilization of wide range of fuels, poly-generation capability, plant flexibility, lower environmental impact, etc.

The aim of this paper was to propose and evaluate conceptual designs of large scale coal gasification plants with pre- and post-combustion capture based on various chemical looping options. Hydrogen and power co-generation was evaluated as potential way to increase energy efficiency of such plants. The plant concepts generated around 420–600 MW net electricity with at least 90% carbon capture rate. For co-generation scenario, a flexible hydrogen output was evaluated in the range of 0–200 MW_{th} (LHV). The results showed net electrical efficiency ranging from 35 to 41%, most of the cases having an almost total carbon capture rate (>99%). Hydrogen co-production show as promising way for efficiency increase.

INTRODUCTION

Energy supply at competitive prices, environmental protection and climate change prevention by reducing greenhouse gas emissions are one of the main issues that modern society is facing. It is known that fossil fuels used in energy sector as well as in other energy-intensive industrial processes are one of the main responsible for greenhouse gas emissions (International Energy Agency, 2012 and Odenberger and Johnsson, 2010). The usage of fossil fuels, especially solid fuels like coal and lignite, is predicted to continue for the years to come (European Commission, 2011). If no significant action is taken to reduce greenhouse gas emissions, severe climatic consequences are predicted (Intergovernmental Panel on Climate Change, 2007). The key to prevent climate change is to reduce anthropogenic greenhouse gas emissions and to develop and deploy new low carbon technologies. Energy efficiency is a key issue in the context of reducing CO_2 emissions. There is a need of strong actions to be taken to put the greenhouse gas emissions in a declining trend for limiting the global temperature increase to 2 °C (European Commission, 2007). In recognition of these aspects, European Union is committed by 2020 to reduce its greenhouse gas emissions by at least 20% compared to 1990 levels together with raising the share of EU energy consumption produced from renewable resources to 20% and a 20% improvement in the EU›s energy efficiency (European Commission, 2008).

Power generation is one of the industrial sectors with major contribution to greenhouse gas emissions (especially CO_2). For climate change mitigation, a special attention is given to the reduction of CO_2 emissions by applying capture and storage techniques in which CO_2 is captured from energy sector and other energy-intensive industrial processes (e.g. cement, metallurgy, petro-chemical, etc.) and then stored in suitable safe geologic locations (Metz et al., 2005). There are several technological options in which CO_2 resulted from fossil fuel-based energy conversion processes is captured (Metz et al., 2005, Davison, 2007, Figueroa et al., 2008, Gibbins and Chalmers, 2008 and Pires et al., 2011).

The most technological and commercially mature carbon dioxide capture method is based on gas–liquid absorption in post-combustion capture configuration. In this case the carbon capture unit is added to remove CO_2 from flue gases before they are vented into atmosphere. CO_2 is removed from the flue gases by gas–liquid absorption into a chemical solvent (e.g. alkanolamines) at relatively low temperatures (40–50 °C). The loaded solvent is then regenerated in a separate column by stripping CO_2. The main drawback of this technology is the significant thermal energy needed to regenerate the solvent, usually in the range of 3–4 MJ/kg CO_2 which in the end implies an energy penalty of about 10 net electricity percentage points for carbon capture (Padurean et al., 2011).

An alternative post-combustion capture option to the classic gas–liquid absorption which was evaluated in this paper referred to the usage of calcium-based carbonation–decarbonation cycle as a possibility to reduce the energy penalty (Fan, 2010, Dean et al., 2011 and Cuadrat et al., 2012). Calcium looping cycle for post-combustion capture implies that the flue gases are contacted with calcium oxide which reacts with carbon dioxide to form calcium carbonate (carbonation reactor). This reactor is operated at 500–650 °C. The two phases are then separated, the clean gas is vented into atmosphere and the solid phase is sent to a calcination reactor where calcium carbonate is decomposed into calcium oxide and carbon dioxide. Both carbonation and calcination reactors are operated in fluidized conditions. The calcination reactor is operated at 900–950 °C. The calcination process is endothermic and a heat source has to be used (usually fuel burning). In order not to dilute the captured CO_2 stream, an oxygen stream (produced by an air separation unit) is used. The chemical reactions of the calcium looping cycle are the following:

$$CO_{2(g)} + CaO_{(s)} \rightarrow CaCO_{3(s)}, \quad \Delta H = -178\,kJ/mol$$

$$(1)$$

$$CaCO_{3(s)} \rightarrow CaO_{(s)} + CO_{2(g)}$$

$$(2)$$

Pre-combustion capture is another capture configuration in which the fuel is partially oxidized with oxygen and steam to produce a mixture of carbon monoxide and hydrogen (called syngas). Gasification is used for solid fuels (coal, lignite, biomass, etc.) and steam catalytic reforming for natural gas (Higman and van der Burgt, 2008). The syngas is then catalytically shifted with steam. The water gas shift (WGS) reaction converts CO into CO_2 and H_2 as follow:

$$CO_{(g)} + H_2O_{(g)} \leftrightarrow CO_{2(g)} + H_{2(g)}$$

$$(3)$$

After WGS conversion, CO_2 can be separated by gas–liquid absorption using either chemical or physical solvents. Pre-combustion capture using gas–liquid absorption show lower energy penalty compared with post-combustion capture due to the fact that CO_2 partial pressure from the treated gas is significantly higher than for post-combustion case. On the power plant level, carbon capture energy penalty in these cases are about 8–9 net electricity percentage points (Davison, 2007, Padurean et al., 2012 and Cormos, 2012a).

This paper proposed four innovative chemical looping-based pre- and post-combustion capture options to be fitted in a gasification plant. All chemical looping configurations imply a continuous double reactor system, a fuel reactor in which the syngas is oxidized to carbon dioxide and water with an oxygen carrier and a steam reactor in which the reduced form of the oxygen carrier is reacting with steam to regenerated the oxygen carrier and produce hydrogen. Both fuel and steam reactors are operated in fluidized mode, the solid being transported by gas phase (Fan et al., 2008, Fan, 2010 and Lyngfelt, 2013).

The first option is based on sorption enhanced water gas shift – SEWGS (Adanez et al., 2012, Cormos and Cormos, 2013 and Lu et al., 2013). In this configuration, carbon monoxide is reacted with water to give carbon dioxide and hydrogen (water gas shift reaction) – see reaction (3). The shift reaction is coupled with CO_2 adsorption on calcium oxide, the overall reaction which take place being:

$$CO_{(g)} + H_2O_{(g)} + CaO_{(s)} \rightarrow CaCO_{3(s)} + H_{2(g)}$$

$$(4)$$

Coupling water shift reaction with CO_2 fixation with calcium oxide has a positive effect on shift reaction equilibrium and carbon monoxide conversion. Formed calcium carbonate is then calcinate in a separate fluidized bed reactor according to reaction (2). As in the post-combustion option presented above, extra fuel has to be combusted with oxygen in the calcination reactor to cover the reaction duty.

The second pre-combustion capture is using iron oxide cycle (Fan, 2010, Cormos, 2012b and Rydén and Arjmand, 2012). In this cycle, the syngas is oxidized in the fuel reactor operated in fluidized bed mode at 600–800 °C with iron oxide (hematite) according to the following reactions:

$$Fe_2O_{3(s)} + 3CO_{(g)} \rightarrow 2Fe_{(s)} + 3CO_{2(g)}$$

$$(5)$$

$$Fe_2O_{3(s)} + 3H_{2(g)} \rightarrow 2Fe_{(s)} + 3H_2O_{(g)}$$

$$(6)$$

The methane present in the syngas is also oxidized to carbon dioxide and water according to the following chemical reaction:

$$4Fe_2O_{3(s)} + 3CH_{4(g)} \rightarrow 8Fe_{(s)} + 3CO_{2(g)} + 6H_2O_{(g)}$$

(7)

The reduced form of the oxygen carrier (iron) is oxidized back in the steam reactor using steam to regenerate the iron oxide and to produce hydrogen according to the reaction:

$$3Fe_{(s)} + 4H_2O_{(g)} \rightarrow Fe_3O_{4(s)} + 4H_{2(g)}$$

(8)

In addition, an air reactor is used to fully oxidize back the iron sorbent to hematite. This process is highly exothermic and besides regenerating the oxygen carrier it is also used to maintain the thermal balance of the whole looping cycle (reactions occurring in the fuel reactor are endothermic).

The third pre-combustion capture option is based on direct coal chemical looping (Fan et al., 2008 and Adanez et al., 2012) in which the fuel was oxidized with iron oxide (hematite) according to the following reaction:

$$Fe_2O_3 + Coal(C_xH_yO_zN_mS_n)_{(s)} \rightarrow Fe_{(s)}$$
$$+ FeS_{(s)} + CO_{2(g)} + H_2O_{(g)} + N_{2(g)}$$

(9)

The reduced form of the oxygen carrier is then reoxidized back is the steam reactor according to the reaction (8). The last case does not use a separate gasifier (as it is the case for other designs) to convert the coal to syngas, the fuel reactor having also the function of converting the solid fuel used.

Both calcium and iron chemical looping cycles are operating at high temperature which enable efficient high temperature heat recovery. This aspect has a positive effect in increasing the energy efficiency of the plant compared with gas–liquid capture processes in which the gas to be treated for CO_2 capture has to be cooled down to 30–40 °C prior to capture. Accordingly, the chemical looping cycles thermal integration in the rest of the gasification plant is an aspect of paramount importance which was treated in detail in the paper (Jerndal et al., 2006 and Budzianowski, 2011).

This paper proposed conceptual designs of gasification plants with pre- and post-combustion capture based on calcium and iron looping cycles. The proposed designs were then evaluated using process flow modelling and process integration techniques. The evaluated plant concepts were considered for power generation but also hydrogen and power co-generation was discussed (Tzimas et al., 2009, Davison, 2011 and Zhang et al., 2013). The analysis was geared towards proposing high energy efficient conceptual designs and then quantify the main plant performance indicators such as: fuel consumption, gross and net energy efficiency, ancillary energy consumption, specific CO_2 emissions, quality specification of captured CO_2 stream, etc. Various plant design aspects (e.g. selection of gasifier, syngas conditioning, chemical looping unit, combined cycle parameters, integration issues, etc.) were investigated in details. As benchmark options, gasification plants without CCS and with CCS based on gas–liquid absorption were considered.

The power plant case studies investigated in the paper produces around 420–600 MW net power with at least 90% carbon capture rate. For poly-generation scenario, the hydrogen and power co-generation was evaluated in a flexible operating case (0–200 MW_{th} hydrogen). The mathematical modelling and simulation of the whole power generation schemes will produce the input data for quantitative technical and environmental evaluations of power plants with carbon capture. Mass and energy integration tools were used to assess the integration aspects of evaluated carbon capture options in the whole power plant design, to optimize the overall energy efficiency and to evaluate the main sources of energy penalty for CCS designs. The results of the evaluations showed that chemical looping systems are very promising as carbon capture options offering an almost total decarbonization rate (in some designs the carbon capture rate was higher that 99%) of the fossil fuel at significantly lower energy penalties compared with conventional CO_2 capture based on gas–liquid absorption (carbon capture energy penalty around 7–8 net electricity percentage points for chemical looping vs. about 10 percentage points for gas–liquid absorption).

PLANT CONFIGURATIONS OF CHEMICAL LOOPING SYSTEMS APPLIED TO GASIFICATION

The conceptual designs investigated in this paper are focused on hydrogen and power co-generation based on coal gasification (Cormos, 2012b and Sorgenfrei and Tsatsaronis, 2013). In the gasifier, the coal is oxidized partially with oxygen and steam to produce syngas. As illustrative gasifier used in this paper, an entrained-flow gasifier with dry fed and gas quench was considered. The choice of this gasifier was based on the fact that this reactor has a high cold gas efficiency (above 80%) and the syngas is free of pyrolysis products which could create problems for the chemical looping unit (Higman and van der Burgt, 2008). The high pressure steam generated in the gasifier island (syngas cooling) is integrated with steam cycle of the combined cycle. Syngas is then desulphurized in an Acid Gas Removal (AGR) system in which H_2S is captured from syngas and send to a Claus plant to be partially oxidized to sulphur. Desulphurized syngas is then used in chemical looping systems to produce hydrogen simultaneous with capturing the carbon from the feedstock.

The first evaluated design (Case 1) is using a calcium-based chemical looping system in post-combustion capture configuration. Basically, this case is similar with an IGCC power plant with addition of the calcium looping unit to treat the gas turbine flue gases. This cases, different from other evaluated cases, is having a syngas-fuelled gas turbine. The hydrogen co-generation capability of this case is limited due to this situation. The hydrogen output is obtained by processing a separate syngas stream in conventional manner (water shift reaction followed Pressure Swing Adsorption –

PSA). The plant layout of this case designed for power generation is presented in Fig. 1.

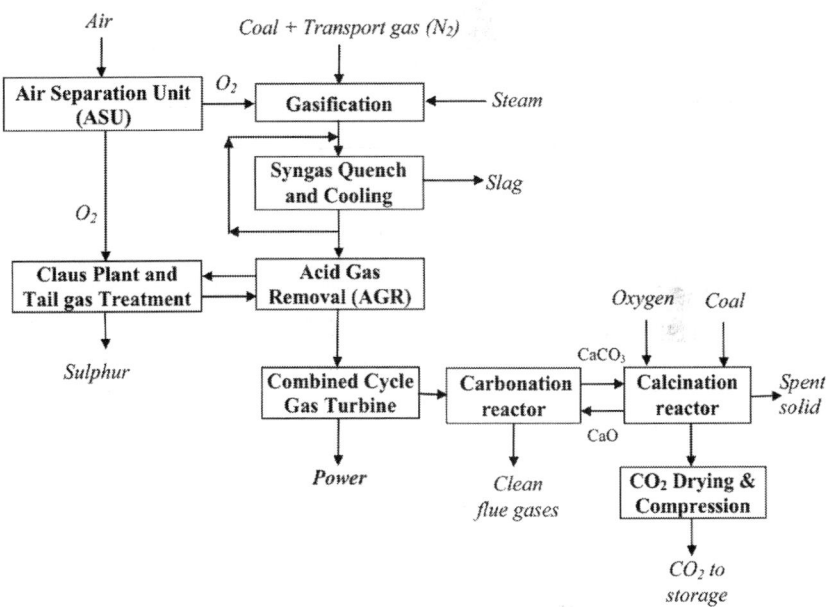

Figure 1: Post-combustion calcium-based looping cycle for gasification plant (Case 1).

The second evaluated option (Case 2) is based also on calcium looping cycle but in a pre-combustion capture configuration. As presented in Section 1 of the paper, the water gas shift reaction was coupled with CO_2 adsorption on calcium oxide (SEWGS). The looping unit has the function to shift the carbon monoxide followed by decarbonize the syngas, the gas turbine of the power block being run on hydrogen. The plant layout is presented in Fig. 2.

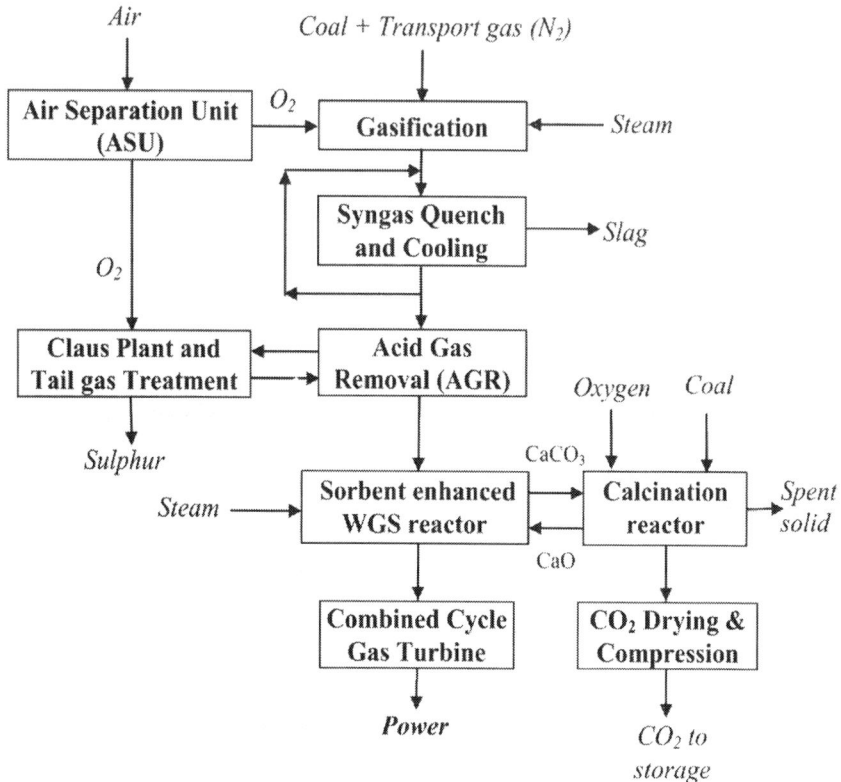

Figure 2: Pre-combustion calcium-based looping cycle for gasification plant (Case 2).

The third option (Case 3) is based on iron looping cycle for decarbonizing the syngas resulted from coal gasification. The hydrogen stream produced in the steam reactor is used then for power production with, if needed, a separate stream which is sent to external customers. The gas turbine of this option is running on hydrogen. The plant layout of this case is presented in Fig. 3.

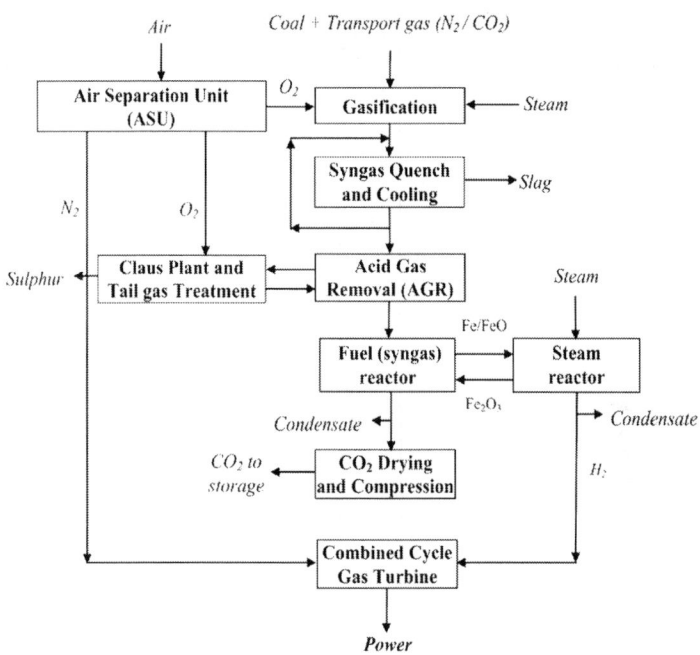

Figure 3: Pre-combustion iron-based looping cycle for gasification plant (Case 3).

The forth option (Case 4) is also based on iron looping cycle but used in direct coal conversion configuration. The plant layout of this case is presented in Fig. 4. As can be noticed from Fig. 4, the fuel reactor there is a combination of gasification (some additional oxygen stream is used apart of the oxygen coming from oxygen carrier) and chemical looping processes. As for Cases 2 and 3, the looping unit is totally decarbonizing the fuel used. The hydrogen stream produced in the steam reactor could be used for power production only or for hydrogen and power co-generation. The gas turbine of this option, as for the Cases 2 and 3, is running on hydrogen. For covering the thermal duties of the reactions in the fuel reactor (endothermic reactions), an additional air reactor is used in combination with steam reactor for reoxidizing the oxygen carrier (exothermic reactions).

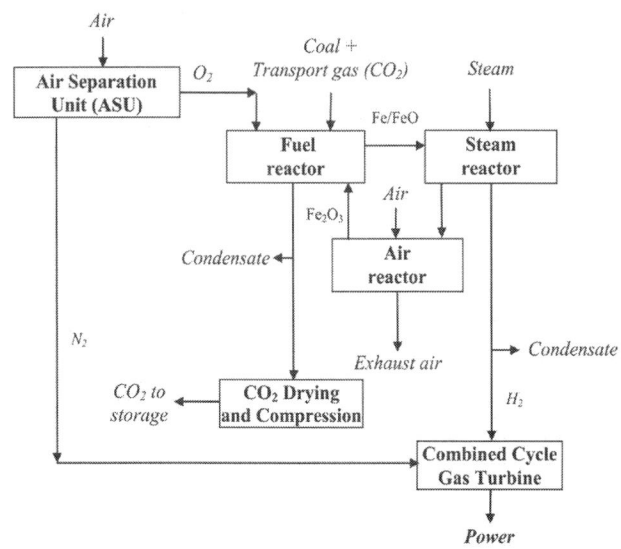

Figure 4: Direct coal chemical looping (Case 4).

For Case 4 (direct coal chemical looping case), the sulphur content in the fuel is transformed in the fuel reactor into iron sulphide which in the air reactor is oxidized to sulphur oxides. The sulphur oxides released with exhaust air are then removed using a limestone-based FGD unit as in classic coal-based power plants (Integrated Pollution Prevention and Control, 2006).

If hydrogen and power co-generation is needed, for all cases, the gas turbine is gradually turned down to 80% to displace an hydrogen stream than can be sent to external customers. The evaluated hydrogen and power co-generation cases are considering a flexible hydrogen output up to 200 MW_{th} based on lower heating value (LHV). An important issue in chemical looping system is the captured CO_2 specification. Most of CO_2 specifications are considering both transport and storage limitations, the general agreement is that utilization of CO_2 for Enhanced Oil Recovery (EOR) implies the most restricted conditions (De Visser et al., 2008). This paper consider this storage option with the following specification (expressed in % vol.): min. 95% CO_2; max 4% non-condensable gases (e.g. nitrogen, argon, hydrogen, etc.); max. 2000 ppm CO; max. 250 ppm water; max. 100 ppm H_2S and max. 10 ppm oxygen. Due to the possible contamination of CO_2 with other gases (the most important being nitrogen used for fuel pneumatic transport as well as argon from oxygen used), for Cases 3 and 4, as coal transport gas two options were evaluated. The first option is using nitrogen (standard gas for gasification plants), the second option being part of captured CO_2. Referring to oxygen stream used in the gasifier, this is delivered by an Air Separation Unit (ASU) with a purity of 95% oxygen, 3% argon and 2% nitrogen (vol.).

As benchmark options used to compare the plant designs using chemical looping systems, one gasification-based power plant without carbon capture was considered (noted Case 5). Also, a gasification plant with CCS based on gas–liquid absorption using Selexol® process was considered (Case 6). Both benchmark options are using the same fuel and the same gasifier (oxygen-blown entrained flow with dry fed and syngas quench) as the chemical looping cases (Cormos, 2012a).

DESIGN ASSUMPTIONS, SIMULATION AND INTEGRATION ASPECTS

All evaluated cases as well as benchmark options were modelled and simulated with ChemCAD and Thermoflex software. The cases were assessed in a common methodology. The composition and thermal characteristics of the evaluated coal sort is presented in Table 1. As main design assumptions, all plant concepts use a common gas turbine (MHI M701G2). This turbine has significant industrial experience in both syngas and hydrogen operation scenario which makes it suitable for designs presented in this paper (Lee et al., 2010). As illustrative example, the description of main plant sub-systems for Case 3 (iron-based looping cycle using syngas) and theirs design assumptions used in the modelling are presented in Table 2(Kruggel-Emden et al., 2011, Cormos, 2012b and Connell et al., 2013).

Table 1: Coal composition and thermal properties

Parameter	Value
Proximate analysis (% wt.)	
Moisture (a.r.)	8.10
Volatile matter (dry)	28.51
Ultimate analysis (% wt. dry)	
Carbon	72.04
Hydrogen	4.08
Nitrogen	1.67
Oxygen	7.36
Sulphur	0.65
Chlorine	0.01
Ash	14.19
Calorific value (kJ/kg)	
Gross (HHV)	28,704.40 (dry)
Net (LHV)	27,803.29 (dry)

Table 2: Main design assumptions (Case 3 – iron-based looping cycle using syngas)

Unit	Parameters
Air separation unit (ASU)	Oxygen purity: 95% (vol.)Power consumption: 225 kWh/ton O_2No air integration with gas turbine (GT)
Gasification reactor (dry fed and syngas quench entrained-flow gasifier)	Oxygen/solid fuel ratio (kg/kg): 0.84Steam/solid fuel ratio (kg/kg): 0.11Nitrogen/solid fuel ratio (kg/kg): 0.09O_2 pressure to gasifier: 48 barSteam pressure to gasifier: 41 barGasification pressure: 40 barGasification temperature: >1400 °C (slagging conditions) Carbon conversion: 99.9%Pressure drop: 1.5 barGas quench type

Syngas quench and conditioning	Syngas temperature after gas quench: ~810 °CTemperature of the quench gas: 250 °CQuench gas ratio: 60%Quench gas compressor efficiency: 80%Pressure drop for fly ash removal system: 1 barHP steam raised in gasification island: 120 bar/375 °CLP steam raised in the gasification island: 3 bar/225 °CHeat exchanger pressure drop: 1% of inlet pressureSyngas temperature after gas boiler: 250 °C
COS hydrolysis	One catalytic bedReactor thermal mode: adiabaticPressure drop: 1 bar
Acid Gas Removal (AGR) unit for H_2S capture	Solvent: Selexol® (dimethyl ethers of polyethylene glycol)H_2S absorption/desorption columns: 24 stages/10 stagesOverall H_2S removal yield: 99.5–99.9%
Claus plant and tail gas treatment	Oxygen-blown typeH_2S recovery: >99%
Chemical looping (CL) unit	Chemical looping agent: magnetite (Fe_3O_4)Fuel reactor parameters: 30.5 bar/750–900 °CSteam reactor parameters: 29.5 bar/500–700 °CGibbs free energy minimization model for both reactorsPressure drop fuel and steam reactors: 1 bar/reactorCL unit fully thermally integrated with the rest of the plant
CO_2 compression and drying	Delivery pressure: 120 barCompressor efficiency: 85%Solvent used for drying: TEG (Tri-Ethylene-Glycol)
Hydrogen compression	Delivery pressure: 60 barCompressor efficiency: 85%
Gas turbine (GT)	Type: M701G2 (Mitsubishi Heavy Industries Ltd.) GT number: 1Net power output: 334 MWElectrical efficiency: 39.5%Pressure ratio: 21Turbine outlet temperature (TOT): 588 °C
Heat Recovery Steam Generator (HRSG) and steam cycle (Rankine)	Three pressure levels (HP/MP/LP): 118/34/3 barMP steam reheatondensation pressure: 0.046 barIntegration of steam generated in gasification island, syngas treatment line and chemical looping unit with CC-GTSteam turbine isoentropic efficiency: 85%Steam wetness ex. steam turbine: max. 10%
Heat exchangers	ΔT_{min} = 10 °CPressure drop: 1% of inlet pressure

The plant concepts were designed with a flexible output of 0–200 MW_{th} hydrogen (based on hydrogen lower heating value – 10.795 MJ/Nm³) with an almost total decarbonization rate of the coal used (>99%). The purity of the hydrogen stream is suitable for usage in chemical and petro-chemical applications as well as other energy conversion processes (e.g. PEM fuel cells for transport sector), which imply very strict quality specification (>99.95% H_2 and virtually no CO and H_2S) due to the possibility of fuel cells poisoning. Captured carbon dioxide stream will have to comply with quality specification presented in Section 2 of the paper. For water removal (lower than 250 ppm), a dehydration step using tri-ethylene-glycol (TEG) was considered. To cover the pressure drop across the transport line and injection requirements, a multi-stage compression with inter-coolers at 120 bar was considered.

As thermodynamic assumption used in the simulations, thermodynamic equilibrium has being assumed for calculations (e.g. gasification, chemical looping, etc.). The choice of thermodynamic equilibrium was considered taking into account the high operating temperatures. Stream property evaluations are based on the Soave–Redlich–Kwong (SRK) equation of state with Boston-Mathias modifications (ChemCAD, 2013) since this takes into account the present chemical species and process operating conditions (pressure, temperature, etc.). For drying unit of captured CO_2 stream, TEG Dehydration thermodynamic package was used. Regarding the chemical looping unit (fuel, steam and air reactors), chemical and phase equilibrium based on a Gibbs free energy minimization model was used in the simulations. Base on available literature information (Chiesa et al., 2008, Fan, 2010, Azimi et al., 2012, Singh et al., 2012, Esmaili et al., 2013 and Lisbona et al., 2013) and the optimized simulation results, the simulation results are in line with experimental results. As illustrative example for Case 3 (iron-based chemical looping using syngas), Table 3 and Table 4 present an in-depth analysis (flowrate, temperature, pressure and composition) of main streams in various plant locations (Table 3) and also focusing on looping reactors (Table 4).

Table 3: Characterization of main plant streams (Case 3 – iron-based looping cycle using syngas)

Stream	Coal	Oxygen (gasifier)	Steam (gasifier)	Nitrogen (gasifier)	Raw syn-gas	Syngas ex. AGR	Captured CO2	Hydrogen (ex. CL)	Flue gas (ex. GT)
Pressure (bar)	Ambient	48.00	41.00	40.00	38.50	31.50	120.00	26.6	1.15
Temperature (°C)	Ambient	80.00	425.00	80.00	1421.14	30.00	35.00	30.00	590.05
Mass flow (kg/h)	162,340	135,846	19,200	14,600	310,865.8	292,717.6	416,610.9	24,750.35	2,687,296
Molar flow (kmole/h)		4224.36	1065.77	521.16	14,580.48	13,594.60	9730.75	12,176.79	98,614.88
Composition (% vol.)									
H_2					26.20	28.09	<0.01	99.90	0.00
CO					57.20	61.34	<0.01	0.00	0.00
CO_2					4.12	4.47	91.54	0.00	0.02
N_2		2		100	4.76	5.11	7.13	0.00	74.42
O_2		95			0.00	0.00	0.00	0.00	11.53
Ar		3			0.87	0.93	1.30	0.00	0.78
H_2S + COS					0.21	5 ppm	6 ppm	0.00	0.00
H_2O			100		6.60	0.02	15 ppm	0.10	13.22
Other					0.04	0.04	0.01	0.00	0.03

Table 4: Characterization of main chemical looping streams (Case 3)

Stream	Syngas reactor		Steam reactor	
	Inlet	Outlet	Inlet	Outlet
Pressure (bar)	31.5	30.5	29.5	28.5
Temperature (°C)	802.25	821.45	700.01	800.00
Mass flow (kg/h)	956,896.2	956,896.2	693,224.6	693,224.6
Molar flow (kmole/h)	17,806.6	21,865.2	20,683.98	16,629.3
Composition (% vol.)				
H_2	21.44	0.03	0.00	73.14
CO	46.83	0.02	0.00	0.00
CO_2	3.41	40.92	0.00	0.00
N_2	3.90	3.17	0.00	0.00
O_2	0.00	0.00	0.00	0.00
Ar	0.71	0.58	0.00	0.00
H_2S	0.00	0.00	0.00	0.00
H_2O	0.02	17.49	60.03	1.52
Iron/Iron oxides	23.65	37.78	39.96	25.32
Other	0.04	0.01	0.01	0.02

All plant concepts analyzed in the paper are modelled and stimulated in a fully thermally integrated design, which means that all the heating duties needed for various processes (e.g. gasifier, chemical looping unit) are based on available hot streams within the plant (e.g. the hot effluents from the fuel, steam and air reactors, hot gas turbine effluent, etc.). The only energy input of the plant being the coal feedstock. The steam flows generated in gasification island, syngas conditioning and chemical looping were integrated in steam cycle of the power block. Pinch analysis was used as main heat and power integration analysis tool (Smith, 2005, Jerndal et al., 2006, Cormos, 2010, Harkin et al., 2010, Bandyopadhyay and Sahu, 2011 and Wang et al., 2013).

For energy integration analysis, the plant was split is two main sub-systems, one being the gasifier island, syngas conditioning line and chemical looping cycle and other being the power block (Combined Cycle Gas Turbine – CCGT). As minimum approach temperature used in pinch analysis, a conservative value of 10 °C was chosen. As illustrative example for Case 3 (iron-based looping using syngas), hot and cold composite curves (HCC and CCC) are presented in Fig. 5 and Fig. 6. Fig. 5 presents hot and cold composite curves for gasifier island, syngas conditioning and chemical looping unit and Fig. 6 presents hot and cold composite curves for the hydrogen-fuelled combined cycle.

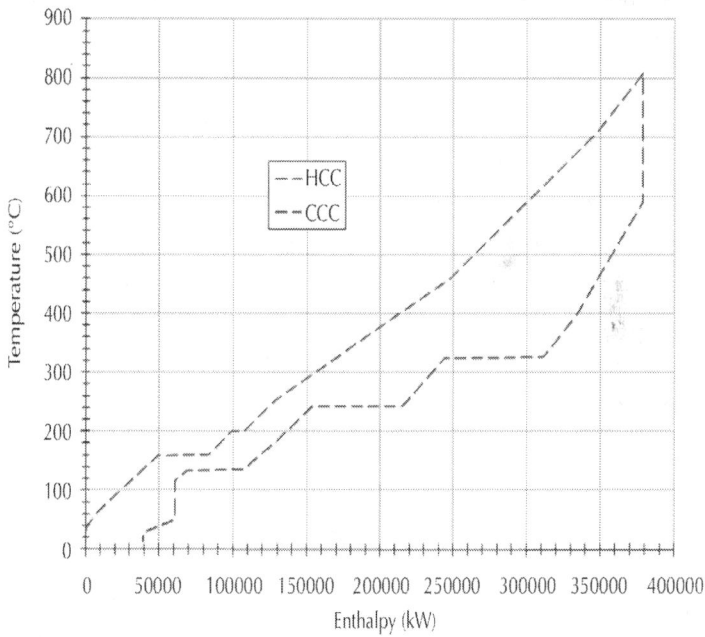

Figure 5: Hot and cold composite curves for Case 3: gasifier island, syngas conditioning and chemical looping unit.

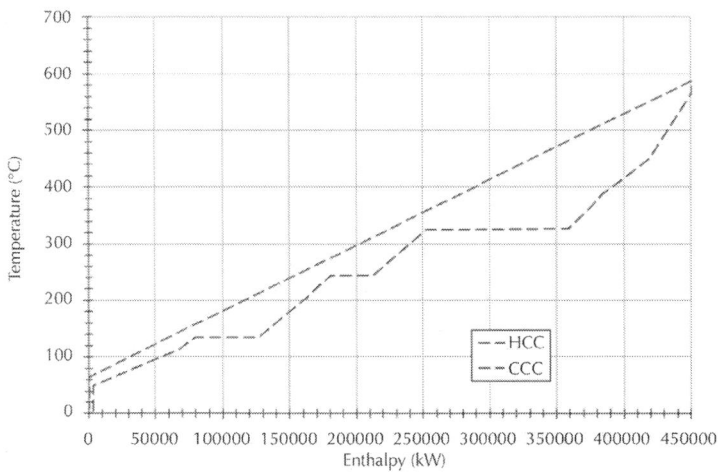

Figure 6: Hot and cold composite curves for Case 3: combined cycle gas turbine.

As can be noticed from Fig. 5 and Fig. 6, the energy flows within the plant were optimized for maximization of plant energy efficiency (Varghese and Bandyopadhyay, 2007 and Cormos, 2012b). The composite curves for gasifier island, syngas conditioning and chemical looping unit are particular for each case. Referring to the composite curves for the power block for other evaluated cases, these are looking very similar with the one presented in Fig. 6 due to the fact that the same gas turbine was used in all cases. This implies the same thermal power of the combustible gas (either syngas for Case 1 or hydrogen for Cases 2–4) going to the gas turbine.

EVALUATION OF CHEMICAL LOOPING-BASED CONCEPTUAL DESIGNS

Simulations of various plant design for hydrogen and power co-generation based on various pre- and post-combustion capture chemical looping cycles yield all necessary process data (mass and molar flows, composition, temperatures, pressures, power generated and consumed) that are needed to assess the overall performance of the processes. The simulation results were compared with experimental data for model validation. No significant differences were found between the simulation results and the experimental results (Fan, 2010, Azimi et al., 2012, Broda et al., 2012 and Singh et al., 2012). Although significant scale-up issues are to be solve before chemical looping technologies become commercial (the current pilot scale are in the range of 1–30 MW thermal), the validation of the mathematical models were done using the existing available information.

As mentioned before, the plant flexibility aspect (capability to adjust plant output in accordance with grid demand) was solved by gradually turn down the gas turbine to make available a hydrogen stream that can be sent to the external customers. At the beginning, all cases were evaluated in power only generation mode. This operation is more likely to be feasible in short to medium term until large scale hydrogen applications are become widely available. The co-generation scenario is a very promising option of plant operation in order to increase the overall energy efficiency. Also, this operation mode are interested to be exploited for plant cycling, the power demand not being constant during time, hydrogen can be co-produce and stored to be used in peak time with profitable economic results.

Table 5 presents the key performance indicators for all evaluated cases operated in power generation only including the two benchmark options (gasification plant without CCS and with CCS based on Selexol® gas–liquid absorption process). As can be noticed from Table 5, the evaluated chemical looping case studies in pre- and post-combustion capture configurations generate about 420–600 MW with net plant electrical efficiencies in the range of 34–41%, the carbon capture rate higher than 95% and specific CO_2 emissions in the range of 3–38 kg/MWh. Some significant differences among various evaluated options can be observed from the plant results.

Table 5: Overall plant performance indicators (power generation only)

Main plant data	Units	Case 1	Case 2	Case 3	Case 4	Case 5	Case 6
Coal flowrate	t/h	226.50	226.71	162.34	145.63	152.50	165.70
Coal LHV	MJ/kg	25.353					
Feedstock thermal energy – LHV (A)	MW_{th}	1595.12	1596.64	1143.28	1025.60	1073.98	1166.98
Gas turbine output	MW_e	334.00	334.00	334.00	334.00	334.00	334.00
Steam turbine output	MW_e	384.60	410.49	199.45	153.49	225.37	210.84
Expander power output	MW_e	1.78	1.40	1.50	22.39	1.78	0.78
Gross electric power output (B)	MW_e	720.38	745.89	534.95	509.88	561.15	545.62
ASU power consumption	MW_e	77.57	70.74	43.82	21.64	41.16	44.73
Gasification island power consumption	MW_e	12.92	9.68	15.06	-	12.92	9.12
AGR + CL + CO_2 drying & compression	MW_e	59.68	52.35	15.18	45.13	10.09	39.81

Parameter	Unit						
Power island power consumption	MW_e	24.24	21.97	22.00	21.72	20.46	18.78
Total ancillary power consumption (C)	MW_e	174.41	154.74	96.06	88.49	84.63	112.44
Net electric power output (D = B – C)	MW_e	545.97	591.15	438.89	421.39	476.52	433.18
Gross electrical efficiency (B/A × 100)	%	45.16	46.71	46.79	49.71	52.24	46.75
Net electrical efficiency (D/A × 100)	%	34.22	37.02	38.38	41.08	44.36	37.11
Carbon capture rate	%	96.07	95.94	99.55	99.34	0.00	90.79
CO_2 specific emissions	kg/MWh	38.47	32.89	3.08	4.13	779.04	86.92

Comparing the results of chemical looping cases, the iron-based designs are performing slightly better than calcium-based designs (both in terms of higher net electrical efficiencies and carbon capture rate). The comparison of pre- and post-combustion capture using chemical looping shows a similar situation as in the case of gas–liquid absorption, the pre-combustion capture cases being more efficient than correspondent post-combustion cases. This is mainly explained by the fact that in pre-combustion capture CO_2 partial pressure is significantly higher than in post-combustion. Pre-combustion capture calcium-looping sorbent enhanced design (Case 2) is closing partially the gap compared with similar design using iron oxide as oxygen carrier but the carbon capture rate is remaining a bit lower and similar with post-combustion case (Case 1).

Comparing the evaluated designs with the benchmark options (Cases 5 and 6), one can notice a mixed picture. The calcium-based post-combustion capture option (Case 1) exhibits the worst performance not only in comparison with other chemical looping designs but also in comparison with carbon capture design based on gas–liquid absorption (Case 6). Case 2 is similar with carbon capture design based on gas–liquid absorption with a bonus in term of carbon capture rate (96% vs. 91%). Both iron-based cases are showing improved performances both in terms of energy efficiency and carbon capture rate (almost total decarbonization of the fuel used). The direct coal chemical looping design (Case 4) is by far the best design, the carbon capture energy penalty of this case being in the range of 3.2 net electricity percentage points, on condition of a close total decarbonization rate.

The second plant operation scenario is based on flexible hydrogen and power co-generation. The power plant flexibility aspects are important in the actual context of modern energy conversion systems in which the share of renewable energy sources (which are highly time irregular) is increasing. This will put an operational burden on fossil fuel-based power plants. As evaluated in this paper, the plant flexibility means the plant core will be operated full load most of the time and only the power block will be operated accordingly to the instant grid demand. This operation mode will have important benefits in term of plant life and finally on economic indicators. This paper evaluates the plant flexibility in the range of in range 0–200 MW$_{th}$ hydrogen output. In this operation range, the gas turbine can be gradually turned down to make available a hydrogen gas stream for external customers.

Table 6 presents the variation of plant performance indicators with hydrogen output (in the range of 0–200 MW$_{th}$ hydrogen) for Case 3 (pre-combustion capture using iron-based cycle with syngas). Several significant aspects can be noticed from the results: the overall plant energy efficiency is increasing, ancillary power consumption and specific CO_2 emissions are decreasing with hydrogen output. These aspects reflect the positive influence of plant flexibility on plant technical parameters. As shown in another work for the case of carbon capture designs using gas–liquid absorption (Cormos, 2012a), this will also have a positive influence on plant economic indicators.

Table 6: Overall plant performance indicators (hydrogen and power co-production mode for Case 3)

Main plant data	Units	Power	Power + hydrogen			
Coal flowrate (a.r.)	kg/h	162.34				
Coal LHV (a.r.)	MJ/kg	25.353				
Feedstock thermal energy – LHV (A)	MW$_{th}$	1143.28				
Gas turbine output	MW$_e$	334.00	313.54	293.05	272.60	252.15
Steam turbine output	MW$_e$	199.45	188.99	179.71	170.39	161.14
Expander power output	MW$_e$	1.50	1.48	1.46	1.44	1.42
Gross electric power output (B)	MW$_e$	534.95	504.01	474.22	444.43	414.71
Hydrogen output – LHV (C)	MW$_{th}$	0.00	50.00	100.00	150.00	200.00

ASU consumption + O_2 compression	MW_e	43.82	43.82	43.82	43.82	43.82
Gasification island power consumption	MW_e	15.06	15.06	15.06	15.06	15.06
AGR + CO_2 drying and compression	MW_e	15.18	15.18	15.18	15.18	15.18
H_2 compression	MW_e	0.00	0.56	1.13	1.70	2.27
Power island power consumption	MW_e	22.00	20.67	19.33	17.99	16.64
Total ancillary power consumption (D)	MW_e	96.06	95.29	94.52	93.75	92.97
Net electric power output (E = B − D)	MW_e	438.89	408.72	379.70	350.68	321.74
Gross electrical efficiency (B/A × 100)	%	46.79	44.08	41.47	38.87	36.27
Net electrical efficiency (E/A × 100)	%	38.38	35.75	33.21	30.67	28.14
Hydrogen efficiency (C/A × 100)	%	0.00	4.37	8.74	13.12	17.49
Cumulative efficiency (C + E/A × 100)	%	38.44	40.12	41.95	43.79	45.63
Carbon capture rate	%	99.55	99.55	99.55	99.55	99.55
CO_2 specific emissions (power + hydrogen)	kg/ MWh	3.08	2.92	2.79	2.68	2.58

Operation of the gasification plants in poly-generation scenario is promising not only from energy point of view of the energy sector. It can be attractive also for chemical and petro-chemical applications is which various chemicals can be produced based on syngas processing (e.g. hydrogen, methanol, substitute natural gas, ammonia, fertilizers, etc.) with simultaneous carbon capture. Another important point of consideration in the integration of gasification processes with other industrial applications lays in interfacing the chemical looping systems evaluated in this paper with cement production and metallurgy applications. Both these energy intensive industrial applications can utilize the spent solid carriers of the looping cycles, calcium oxide in cement production and iron oxide in metal production (Vatopoulos and Tzimas, 2012).

As mentioned before, an important aspect evaluated in the paper was captured carbon dioxide quality specification. The normal way in which the solid fuel is transported to the fuel reactor is based on pneumatic transport using an inert gas. Usually in an gasification power plant nitrogen is used for this purpose. For chemical looping systems, the usage of nitrogen as inert gas to transport the solid fuel to the conversion reactor rises the possibility of impurification of captured CO_2 stream with nitrogen. Captured CO_2 specification has to comply with the transport and storage requirements. In this paper, the more restrictive storage option (namely EOR) was selected. According to quality specification the content of CO_2 has to be at least 95% (vol.). For the chemical looping options evaluated in this paper, the more prone to nitrogen contamination are the iron-based ones in which both fuel nitrogen and nitrogen used for fuel pneumatic transport are ended in the captured CO_2 stream. For calcium-based design the nitrogen contamination is somehow not as acute due to the fact that calcium carbonate is the carbon capture vector and this separate the carbonation reactor from the calcinations reactor. The only nitrogen contamination in these cases is coming from additional coal to be combusted with oxygen in the calcinations reactor.

Table 7 presents the quality specification of captured CO_2 streams of both evaluated iron-based cases using nitrogen and carbon dioxide as inert transport gases. As can be noticed, the usage of nitrogen implies the impurification of captured stream below 95% CO_2 level as required, but the usage of carbon dioxide solves this issues. For calcium-based cases, captured CO_2 stream has a purity of 98–99% vol. even using the nitrogen for fuel transport.

Table 7: Quality specification of captured CO_2 streams using various coal transport gases (nitrogen and carbon dioxide)

Composition (% vol.)	Proposed specification	Case 3 (N_2)	Case 3 (CO_2)	Case 4 (N_2)	Case 4 (CO_2)
CO_2	>95.00	92.51	97.89	92.38	97.59
CO	<2000 ppm	200 ppm	75 ppm	50 ppm	80 ppm
O_2	<10 ppm	0.00	0.00	0.00	0.00
N_2	<4.00 (all non-cond. gases)	6.94	1.58	6.81	1.60
Ar		0.45	0.48	0.77	0.78
H_2		0.07	0.01	0.01	0.01
H_2S + COS	<100 ppm	7 ppm	14 ppm	10 ppm	15 ppm
H_2O	<250 ppm	10 ppm	11 ppm	11 ppm	12 ppm
Other		0.03	0.03	0.01	0.01

CONCLUSIONS

The paper assessed the performance of various chemical looping configurations for hydrogen and power co-generation based on coal gasification process. As evaluated solid sorbents, calcium and iron-based materials were used due their performances, low costs, environmental impact aspects and possibility to integrate with other industrial applications. Both pre- and post-combustion capture designs of looping cycles were assessed. The evaluated cases were compared with two benchmark options: the first option is a design without CCS, the second option implies pre-combustion capture using Selexol® solvent.

All design options were investigated by an integrated modelling and simulation framework, the concepts being also optimized in term of energy efficiency by heat and power integration analysis (pinch method). The mass and energy balances were used to assessed the overall plant performances. The results showed that iron-based direct chemical looping option (Case 4) is by far the most efficient design, the net electrical efficiency is about 41%, an almost total carbon capture rate (>99%) and a carbon capture energy penalty in the range of 3.2 net electricity percentage points. Iron-based design using syngas (Case 3) is also showing good performances, e.g. carbon capture rate higher than 99% and energy penalty in the range of 6%. Referring to calcium-looping options, the pre-combustion concept (Case 2) is performing comparable with benchmark carbon capture option based on gas–liquid absorption. The post-combustion capture design (Case 1) is performing significantly worse in comparison with other chemical looping designs and also benchmark carbon capture option.

One important aspect evaluated in the paper was the plant flexibility (the capability of the plant to adjust the output energy vectors according to instant grid demand). In this respect, hydrogen and power co-generation operation scenario were assessed. As results showed, in co-generation operation the main performance indicators (net energy efficiency, ancillary power consumption, specific CO_2 emissions) are improving with increasing the hydrogen output. For instance the flexible operation scenario in the range from 0 to 200 MWth hydrogen implies an increase of the overall plant energy efficiency with 7 net energy percentage points (cumulative hydrogen and power). The overall conclusion is that the plant flexibility is significantly improving the plant techno-economic indicators.

ACKNOWLEDGMENTS

This work was supported by Romanian-Swiss Research Programme, project no. IZERZ0_141976/1(13 RO-CH/RSRP/2013): "Advanced thermo-chemical looping cycles for the poly-generation of decarbonised energy vectors: Material synthesis and characterisation, process modelling and life cycle analysis".

REFERENCES

1. Adanez, J., Abad, A., Garcia-Labiano, F., Gayan, P., de Diego, L., 2012. Progress in chemical-looping combustion and reforming technologies. Prog. Energy Combust. Sci. 38, 215–282.

2. Azimi, G., Keller, M., Mehdipoor, A., Leion, H., 2012. Experimental evaluation and modeling of steam gasification and hydrogen inhibition in chemical-looping combustion with solid fuel. Int. J. Greenhouse Gas Control 11, 1–10.

3. Bandyopadhyay, S., Sahu, G.C., 2011. Modified problem table algorithm for energy targeting. Ind. Eng. Chem. Res. 49, 11557–11563.

4. Broda, M., Kierzkowska, A., Müller, C., 2012. Influence of the calcination and carbonation conditions on the CO_2 uptake of synthetic Ca-based CO_2 sorbents. Environ. Sci. Technol. 46, 10849–10856.

5. Budzianowski, W., 2011. Low-carbon power generation cycles: the feasibility of CO_2 capture and opportunities for integration. J. Power Technol. 91, 6–13.

6. ChemCAD, 2013. Chemical Process Simulation – Version 6.5. Chemstations, Huston, USA, http://www.chemstations. com (accessed 10.06.13).

7. Chiesa, P., Lozza, G., Malandrino, A., Romano, M., Piccolo, V., 2008. Three-reactors chemical looping process for hydrogen production. Int. J. Hydrogen Energy 33, 2233–2245.

8. Connell, D., Lewandowski, D., Ramkumar, S., Phalak, N., Statnick, R., Fan, L.S., 2013. Process simulation and economic analysis of the calcium looping process (CLP) for hydrogen and electricity production from coal and natural gas. Fuel 105, 383–396.

9. Cormos, C.C., 2010. Evaluation of energy integration aspects for IGCC-based hydrogen and electricity co-production with carbon capture and storage. Int. J. Hydrogen Energy 35, 7485–7497.

10. Cormos, C.C., 2012a. Integrated assessment of IGCC power generation technology with carbon capture and storage (CCS). Energy 42, 434–445.

11. Cormos, C.C., 2012b. Evaluation of syngas-based chemical looping applications for hydrogen and power co-generation with CCS. Int. J. Hydrogen Energy 37, 13371–13386.

12. Cormos, C.C., Cormos, A.M., 2013. Assessment of calcium-based chemical looping options for gasification power plants. Int. J. Hydrogen Energy 38, 2306–2317.

13. Cuadrat, A., Abad, A., de Diego, L., García-Labiano, L., Gayán, P., Adánez, J., 2012. Prompt considerations on the design of chemical-looping combustion of coal from experimental tests. Fuel 97, 219–232.

14. Davison, J., 2007. Performance and costs of power plants with capture and storage of CO2. Energy 32, 1163–1176.

15. Davison, J., 2011. Flexible CCS plants – a key to near-zero emission electricity systems. Energy Procedia 4, 2548–2555.

16. De Visser, E., Hendriks, C., Barrio, M., Mølnvik, M.J., de Koeijer, G., Liljemark, S., 2008. Dynamis CO2 quality recommendations. Int. J. Greenhouse Gas Control 2, 478–484.

17. Dean, C.C., Blamey, J., Florin, H.N., Al-Jeboori, M.J., Fennell, P.S., 2011. The calcium looping cycle for CO2 capture from power generation, cement manufacture and hydrogen production. Chem. Eng. Res. Des. 89, 836–855.

18. Esmaili, E., Mahinpey, N., Jim Lim, C., 2013. Modified equilibrium modelling of coal gasification with in situ CO2 capture using sorbent CaO: assessment of approach temperature. Chem. Eng. Res. Des. (in press).

19. European Commission, 2007. Limiting Global Climate Change to 2 ᵒCelsius – The Way Ahead for 2020 and Beyond. COM(2007) 2 final.

20. European Commission, 2008. Communication from the Commission. 20 20 20 by 2020: Europe's climate change opportunity. COM(2008) 30 final.

21. European Commission, 2011. Communication from the Commission: Energy Roadmap 2050. COM(2011) 885/2.

22. Fan, L.S., Li, F., Ramkumar, S., 2008. Utilization of chemical looping strategy in coal gasification processes. Particuology 6, 131–142.

23. Fan, L.S., 2010. Chemical looping systems for fossil energy conversions. Wiley-AIChE.

24. Figueroa, J.D., Fout, F., Plasynski, S., McIlvired, H., Srivastava, R., 2008. Advances in CO2 capture technology – The U.S. Department of Energy's Carbon Sequestration Program. Int. J. Greenhouse Gas Control 2, 9–20.

25. Gibbins, J., Chalmers, H., 2008. Carbon capture and storage. Energy Policy 36, 4317–4322.

26. Harkin, T., Hoadley, A., Hooper, B., 2010. Reducing the energy penalty of CO2 capture and compression using pinch analysis. J. Clean. Prod. 18, 857–866.

27. Higman, C., van der Burgt, M., 2008. Gasification, second ed. Gulf Professional Publishing, Elsevier Science, Burlington.

28. International Energy Agency (IEA), 2012. Energy Technology Perspective.

29. Intergovernmental Panel on Climate Change (IPCC), 2007. Fourth Assessment Report: Climate Change.

30. Integrated Pollution Prevention and Control (IPPC), 2006. Reference document on Best Available Techniques for Large Combustion Plants. European Commission, JRC, Seville.

31. Jerndal, E., Mattisson, T., Lyngfelt, A., 2006. Thermal analysis of chemical-looping combustion. Chem. Eng. Res. Des. 84, 795–806.

32. Kruggel-Emden, H., Stepanek, F., Munjiza, A., 2011. A comparative study of reaction models applied for chemical looping combustion. Chem. Eng. Res. Des. 89, 2714–2727.

33. Lee, M.C., Seo, S.B., Chung, J.H., Kim, S.M., Joo, Y., Ahn, D.H., 2010. Gas turbine combustion performance test of hydrogen and carbon monoxide synthetic gas. Fuel 89, 1485–1491.

34. Lisbona, P., Martínez, A., Romeo, L., 2013. Hydrodynamical model and experimental results of a calcium looping cycle for CO2 capture. Appl. Energy 101, 317–322.

35. Lu, H., Lu, Y., Rostam-Abadi, M., 2013. CO2 sorbents for a sorption-enhanced water–gas-shift process in IGCC plants: a thermodynamic analysis and process simulation study. Int. J. Hydrogen Energy 38, 6663–6672.

36. Lyngfelt, A., 2013. Chemical-looping combustion of solid fuels – status of development. Appl. Energy (in press).

37. Metz, B., Davidson, O., de Coninck, H., Loos, M., Meyer, L., 2005. Carbon Dioxide Capture and Storage. Intergovernmental Panel on Climate Change (IPCC).

38. Odenberger, M., Johnsson, F., 2010. Pathways for the European electricity supply system to 2050 – the role of CCS to meet stringent CO2 reduction targets. Int. J. Greenhouse Gas Control 4, 327–340.

39. Padurean, A., Cormos, C.C., Cormos, A.M., Agachi, P.S., 2011.

40. Multicriterial analysis of post-combustion carbon dioxide capture using alkanolamines. Int. J. Greenhouse Gas Control 5, 676–685.

41. Padurean, P., Cormos, C.C., Agachi, P.S., 2012. Pre-combustion carbon dioxide capture by gas–liquid absorption for integrated gasification combined cycle power plants. Int. J. Greenhouse Gas Control 7, 1–11.

42. Pires, J.C.M., Martins, F.G., Alvim-Ferraz, M.C.M., Simões, M., 2011.

43. Recent developments on carbon capture and storage: an overview. Chem. Eng. Res. Des. 89, 1446–1460.

44. Rydén, M., Arjmand, M., 2012. Continuous hydrogen production via the steam-iron reaction by chemical looping in a circulating fluidized-bed reactor. Int. J. Hydrogen Energy 37, 4843–4854.

45. Singh, A., Al-Raqom, F., Klausner, J., Petrasch, J., 2012.

Production of hydrogen via an iron/iron oxide looping cycle: thermodynamic modeling and experimental validation. Int. J. Hydrogen Energy 37, 7442–7450.

46.   Smith, R., 2005. Chemical Processes: Design and Integration. Wiley, West Sussex, England.

47.   Sorgenfrei, M., Tsatsaronis, G., 2013. Design and evaluation of an IGCC power plant using iron-based syngas chemical-looping (SCL) combustion. Appl. Energy (in press).

48.   Tzimas, E., Cormos, C.C., Starr, F., Garcia-Cortes, C., 2009. The design of carbon capture IGCC-based plants with hydrogen co-production. Energy Procedia 1, 591–598.

49.   Varghese, J., Bandyopadhyay, S., 2007. Targeting for energy integration of multiple fired heaters. Ind. Eng. Chem. Res. 46, 5631–5644.

50.   Vatopoulos, K., Tzimas, E., 2012. Assessment of CO_2 capture technologies in cement manufacturing process. J. Clean. Prod. 32, 251–261.

51.   Wang, D., Chen, S., Xu, C., Xiang, W., 2013. Energy and exergy analysis of a new hydrogen-fueled power plant based on calcium looping process. Int. J. Hydrogen Energy 38, 5389–5400.

52.   Zhang, X., Gundersen, T., Roussanaly, S., Brunsvold, A., Zhang, S., 2013. Carbon chain analysis on a coal IGCC-CCS system with flexible multi-products. Fuel Process. Technol. 108, 146–153.

Chapter 5

Model Design of a Class of Moving-Bed Tubular Gasification Reactors

Ulises Badillo-Hernandez[a], Luis Alvarez-Icaza[a], and Jesus Alvarez[b,]

[a]Universidad Nacional Autónoma de México, Instituto de Ingeniería, Distrito Federal, C.P. 04510, Mexico
[b]Universidad Autónoma Metropolitana-Iztapalapa, Departamento de Ingeniería de Procesos e Hidráulica, Distrito Federal, C.P. 09340, Mexico

ABSTRACT

The problem of modeling a class of two-phase moving-bed tubular gasification reactors by means of a lumped (finite-dimensional) representation is addressed in this paper. A model is designed, as simple as possible, in the light of a specific – experimental, equipment, operation, monitoring or control – design task and the uncertainty of the underlying kinetics and transport parameters.

First, the enforcement of quasi-steady state (QSS) gas-phase assumptions and stoichiometric considerations followed by spatial finite-difference approximation plus interpolation leads to a representation of the tubular reactor as a train of N continuous stirred tank reactors (CSTRs). Then, the number of tanks and their volumes are chosen according to the modeling objective. The proposed approach is illustrated with a case example, studied before with experiments and PDE-based simulations, finding that the dynamics of the tubular reactor can be modeled with three CSTRs (9-ODE).

INTRODUCTION

The conversion of a carbonaceous solid fuel into a gaseous one (syngas or producer gas) has become an interesting alternative for power generation and energy conversion (Beér, 2007). A diversity of fluidized, entrained and moving-bed two-phase tubular reactors has been employed to carry out the reaction. A state-of-the-art description of the wide variety of gasification technologies is provided in the literature (Reed and Das, 1988, Basu, 2006 and Higman and van der Burgt, 2008). This paper focuses on the moving-bed reactor, which has the advantage of enabling small-scale operations with a suitable trade off between product quality and variability of solid feed composition.

In two extensive studies (Amundson and Arri, 1978 and Caram and Fuentes, 1982), according to the standard chemical reactor modeling plus space finite-differences discretization, it was established that in operation at maximum thermal efficiency a countercurrent moving-bed reactor exhibited the following behavior: (i) strong parametric sensitivity with respect to feed flows, (ii) traveling reaction front behavior, and (iii) three steady-state profile sets for a certain solid flow rate interval. These features were validated in subsequent experimental steady-state (Manurung and Beenackers, 1993, Reed et al., 1999, Zainal et al., 2002 and Sheth and Babu, 2009) and transient (Reed and Markson, 1985, Barrio et al., 2001 and Shwe, 2004) studies. In particular, Reed and Markson (1985) and Barrio et al. (2001) reported that: (i) two different steady-states were possible, one with reaction front close to the reactor solid phase inlet and high solid conversion, and the other one with front close to the exit and low conversion and (ii) either steady-state can be reached depending on the feed flow rate value. In an experimental study of a coal gasifier with PID temperature control (Chen et al., 2007), it was reported an initial transient response dominated by pyrolysis (due to the relatively high yields of H_2 and CH_4) followed by a response governed by char gasification (high measured concentrations of CO and CO_2). In view of Amundson and Arri (1978) study, it is not clear if the open-loop operations are unique or not, and if they were or not close to bifurcation conditions.

Recently, simulation studies have been conducted for performance and parameter sensitivity evaluation of the biomass feed processes (Di Blasi, 2000, DiBlasi, 2004, Rogel and Aguillon, 2006, Gobel et al., 2007 and Grieco and Baldi, 2011) with models that: (i) constitute variations and/or refinements of the one developed by Amundson and Arri (1978) and (ii) yield rather good quantitative descriptions of experimental data. Among the refinements are: more components and reactions (two step pyrolysis finite kinetics), gas-phase quasi-steady-state (QSS) assumption, radial dependency of the profiles, thermal equilibrium between phases and use of computational fluid dynamics (CFD) tools. The models describe the experimental data with 8–15 and 5–12 percent deviations for absolute temperature and gas effluent concentrations, respectively.

Even though these models suffice for process design and redesign, they may be unduly high dimensional and stiff due to the absence of gas-phase QSS assumptions (Gobel et al., 2007), and of stoichiometry-based partition of species (Aris, 1965 and Feinberg, 1977). The availability of reduced order models should: (i) facilitate studies on the key multiplicity and stability issues and (ii) enable the development of tractable advanced monitoring and control schemes, in the understanding that the stability, estimation and control theory for infinite dimensional systems lags quite behind the one for finite dimensional ones. These considerations motivate the scope of the present study: the development of a reduced-order modeling framework for tubular moving-bed gasification reactors.

The methodological assumptions used in this paper are based on the idea that, in constructive control (Sepulchre et al., 1997 and Krstic et al., 1995), the choice of model itself is a design degree of freedom that can be effectively exploited to devise nonlinear observers and (advanced and conventional) controllers for CSTRs (Lopez and Alvarez, 2004, Diaz-Salgado et al., 2012 and Schaum et al., 2012), for drying distributed systems (Martinez-Vera et al., 2010), and (distributed-like) staged systems (Castellanos-Sahagun et al., 2006 and Fernandez et al., 2012).

In this paper, the problem of designing the model, in the light of a specific objective, for a class of two-phase moving-bed tubular gasification reactors is addressed. Among the modeling objectives are: experimental, equipment, operation, monitoring and control design. The problem consists in developing the simplest possible lumped (finite-dimensional) representation of the syngas tubular reactor such that the particular modeling objective is met with the smallest number and weakest coupling of ordinary differential equations (ODEs), relative to the uncertainty of the kinetics and transport parameters and the prescribed model accuracy.

First, the enforcement of quasi-steady state (QSS) gas-phase assumptions and stoichiometric considerations, followed by spatial finite-difference (FD) approximation plus interpolation leads to a reduced order dynamical model of the reactor, that is interpreted as a train of N continuously stirred tank reactors (CSTRs) with backmixing. The number of tanks and their volumes are regarded as design degrees of freedom and chosen according to the modeling objective. The proposed approach is applied to a case example studied previously with experiments (Manurung and Beenackers, 1993) and simulations (Di Blasi, 2000), finding that reactor dynamics can be adequately modeled with three CSTRs (9-ODE).

REACTOR MODELING PROBLEM

In this work, a continuous moving-bed tubular gasifier reactor (depicted in Fig. 1) is considered where solid-to-gas fuel conversion occurs according to a multicomponent reaction network in the gas and solid phases. Two feed streams are connected to the reactor, one at the top with the solid fuel particles (coal, municipal waste or wood pellets, assumed mono-disperse), and the other one supplying the gasification agent (air, oxygen, steam or a mixture of them) either from the top or from the bottom. The reactor is equipped with two exit streams, one with the unreacted solid particles (ash and char), and the other with the syngas product. Without restricting the approach, this paper is circumscribed to the case of downdraft operation with concurrent feed flow pair (Fig. 1).

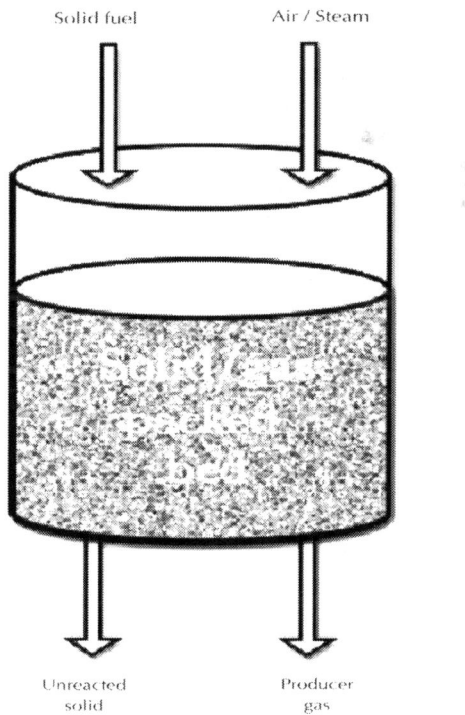

Figure 1: Concurrent moving-bed tubular gasifier.

Reactor Distributed System

For the modeling purpose at hand, let us regard that solid-to-gas fuel conversion occurs through a pseudo-homogeneous reaction network of m reactions, n_π^R reactive components, and n_π^I inert components $n\pi = n_\pi^R + n_\pi^I$ in the ϖ (gas or solid) phase, according to the following expression in stoichiometry-oriented form (Aris, 1965):

$$s_{1,j}^\pi \Omega_1^\pi + \cdots + s_{n_\pi^R j}^\pi \Omega_{n_\pi}^\pi = 0, \quad j = 1, \ldots, m \tag{1a}$$

$$\mathbf{R}_j(\mathbf{A}, T), \quad \Delta H_j, \quad \pi = g, s$$
$$g = gas, \quad s = solid$$
$$\mathbf{A}_\pi = [A_1^\pi, A_2^\pi, \ldots, A_{n_\pi}^\pi]^T \tag{1b}$$

$$\mathbf{A} = [\mathbf{A}_g^T, \mathbf{A}_s^T]^T, \quad \mathbf{T} = [T_s, T_g]^T \tag{1c}$$

where Ω_i^π is the i-th component in the ϖ phase, $S_{i,j}^\pi$ is the stoichiometric coefficient of the i-th component Ω_i^π in the j-th reaction of the ϖ phase, R_j is the j-th reaction rate function, T_ϖ is the temperature of the ϖ phase, H_j is the enthalpy change of the j-th reaction, A_ϖ is the vector with molar concentrations A_π (per unit volume) of the i-th component in the ϖ phase, and $\mathbf{A}$ is the vector with concentrations in both phases.

In terms of the preceding reaction network description, the spatially distributed models developed in recent studies (Di Blasi, 2000, DiBlasi, 2004 and Shwe, 2004) of moving-bed gasifier (depicted in Fig. 1) are given by the following set of partial differential equations (PDEs) over the spatial ($0 \leq Z \leq L$) and temporal ($t' > 0$) domain:

$$\partial_{t'} A_i^S = \Gamma_M^S\{A_i^S\} + \mathcal{G}_i^S(\mathbf{A}, \mathbf{T}), \quad i = 1, \ldots, n_s \tag{2a}$$

$$\partial_{t'} T_s = \Gamma_H^s\{T_s\} + Q_R^s(\mathbf{A}, \mathbf{T}) - Q_{sw} - Q_{sg} \qquad (2b)$$

$$\partial_{t'} A_i^g = \Gamma_M^g\{A_i^g\} + \mathcal{G}_i^g(\mathbf{A}, \mathbf{T}), \quad i = 1, ..., n_g \qquad (2c)$$

$$\partial_{t'} T_g = \Gamma_H^g\{T_g\} + Q_R^\pi(\mathbf{A}, \mathbf{T}) - Q_{gw} + Q_{sg} \qquad (2d)$$

Boundary conditions

$$Z = 0 : B_M^s\{\mathbf{A}_s\} = 0, \qquad B_M^g\{\mathbf{A}_g\} = 0,$$
$$B_H^s\{T_s\} = 0, \qquad B_H^g\{T_g\} = 0 \qquad (2e)$$

$$Z = L : \partial_Z \mathbf{A}_s = 0, \qquad \partial_Z \mathbf{A}_g = 0,$$
$$\partial_Z T_s = 0, \qquad \partial_Z T_g = 0 \qquad (2f)$$

Initial conditions

$$t' = 0 : \mathbf{A}_s(Z, 0) = \mathbf{A}_{s,ic} \; T_s(Z, 0) = T_{s,ic}, \qquad (2g)$$

$$t' = 0 : \mathbf{A}_s(Z, 0) = \mathbf{A}_{s,ic} \; T_s(Z, 0) = T_{s,ic}, \qquad (2h)$$

where

$$\mathcal{G}_i^\pi(\mathbf{A}, \mathbf{T}) = \sum_{j=1}^{m} s_{i,j}^\pi R_j(\mathbf{A}, \mathbf{T}), \quad i = 1, ..., n_\pi^R$$

(3a)

$$Q_R^\pi(\mathbf{A}, \mathbf{T}) = -\sum_{j=1}^{m_\pi} \frac{\Delta H_j}{C_\pi C_p^\pi} R_j(\mathbf{A}, \mathbf{T}), \quad \pi = g, s \qquad (3b)$$

$$Q_{sg} = A_s h_{sg}[T_s - T_g], \qquad (3c)$$

$$Q_{\pi w} = A_v h_{\pi w}[T_\pi - T_w], \quad A_v = \frac{4}{D_R} \qquad (3d)$$

G_i^π is the i -th component rate of generation in the ϖ phase,

G_R^π is the heat rate of generation in the ϖphase, Qsg is the solid–gas interphase heat transfer rate, and $Q_{\varpi g}$ is the ϖ phase-wall heat exchange rate.

Γ_M^π (or Γ_M^π (defined in Table 1) is the mass (or heat) convection–diffusion transport operator for the ϖphase, ug (or us) is the gas (or solid) convective velocity, $_{DM}$ is the mass dispersion coefficient of gas phase, D_M is the heat diffusion coefficient for the ϖ phase, $_{C\varpi}$ is the ϖ-phase total molar concentration, and C_R^π is the specific heat capacity of the ϖ phase (see Table 2 for nomenclature and units of variables). The boundary operators for the mass B_R^π and heat B_H^π balances in the ϖ phase are also defined in Table 1.

Table 1: Operators of the moving-bed reactor model (2a), (2b), (2c), (2d), (2e), (2f), (2g) and (2h)

Boundary operators
$B_M^g\{\mathbf{X}\}:=\mathcal{D}_M\partial_Z\mathbf{X}-u_g[\mathbf{X}-\mathbf{X}_e]$
$B_M^s\{\mathbf{X}\}:=\mathbf{X}-\mathbf{X}_e$
$B_H^\pi\{T_\pi\}:=\mathcal{D}_H^\pi\partial_z T_\pi-u_\pi[T_\pi-T_e]$
Transport operators
$\Gamma_M^g\{\mathbf{X}\}(\mathbf{A},\mathbf{T}):=\partial_z(-u_g(\mathbf{A},\mathbf{T})\,\mathbf{x}+\mathcal{D}_M\partial_z\mathbf{x})$
$\Gamma_M^s\{\mathbf{X}\}(\mathbf{A},\mathbf{T}):=\partial_z(-u_s(\mathbf{A},\mathbf{T})\mathbf{x})$
$\Gamma_H^\pi\{T_\pi\}(\mathbf{A},\mathbf{T}):=\partial_z(-u_\pi(\mathbf{A},\mathbf{T})\mathbf{x}+\mathcal{D}_H^\pi\partial_z T_\pi)$
$\mathcal{D}_H^\pi=\dfrac{\lambda_\pi^{eff}}{C_\pi C_p^\pi}$

Table 2: Nomenclature

A_j^{π}	Molar concentration of i -th component at π phase in kmol/m^3
$T\pi$	Temperature of the π-phase in K
R_j	Rate of j -th reaction in $\text{kmol}/(\text{m}^3\text{ s})$
Δ_{Hj}	Enthalpy change of j -th reaction in kJ/kmol
$u\pi$	Axial flow velocity of π phase in m/s
uH	Total heat flow velocity of both phases in m/s
DM	Gas diffusion coefficient in m^2/s
DH	Heat diffusion coefficient in m^2/s
$C\pi$	Total molar concentration of π phase in kmol/m^3
C_{p}^{π}	Specific heat capacity of π phase in $\text{kJ}/(\text{kmol K})$
χ_{π}^{F}	Effective thermal conductivity of π phase in $\text{kW}/(\text{m K})$
hw	Heat transfer coefficient to/from reactor wall at $\text{kW}/(\text{m}^2\text{ K})$
Av	Heat transfer area between phases and reactor wall at m^{-1}
AR	Transversal internal area of reactor at m^2
DR	Reactor internal diameter in m
L	Height of reactor in m
Pg	Pressure of gas phase in Pa
μg	Viscosity of gas phase in $\text{kg}/(\text{m s})$
W_j^{π}	Molar mass of i component of π phase in kg/kmol
Lj	Arrhenius factor of j-th reaction
Ej	Activation energy of j -th reaction in kJ/(kmol)
U	Universal gas constant in $\text{kJ}/(\text{kmol K})$
ηp	Particle number density in m^{-3}
dp	Particle diameter in m^{-1}
Up	Particle area m^2

X	Char molar fraction
χash	Molar fraction of ash in char particle
dp_0	Initial particle diameter in m^{-1}
km	Mass transfer coefficient between phases in m/s
k_m^*	Maximum mass transfer coefficient in m/ s

Modeling Problem

In recent related moving-bed gasifier studies (Di Blasi, 2000, DiBlasi, 2004, Rogel and Aguillon, 2006 and Gobel et al., 2007), the statics and some aspects of the dynamics have been examined through numerical approximations of model (2a), (2b), (2c), (2d), (2e), (2f), (2g) and (2h) using finite differences, finite volume and CFD packages over a sufficiently dense spatial mesh. The underlying models consist of 100–500 dynamic lumps (internal nodes of spatial mesh) with 1800–7000 ODEs (see Table 7). Eventhough these studies have provided valuable results and insight, further tractable studies on process, control and monitoring strategy design require the development of reduced-order reactor models.

Specifically, the present study focuses in designing a dynamic model with the smallest possible number and the weakest coupling of ODEs in the light of a specific modeling objective: bifurcation, multiplicity and stability analysis of steady states, as well as, process (equipment and operation conditions), control and observer synthesis.

Case Example

For illustration, testing, and comparison purposes of the proposed modeling approach, a representative case example studied before using experiments and simulations (Di Blasi, 2000) is selected: a pilot scale (DR=0.45 m and L=0.5 m) downdraft biomass-air moving-bed gasifier. The biomass to syngas conversion is described by an unreacted shrinking core particle model (Di Blasi, 2000) in terms of the Schmidt (Sc) and the Reynolds (Re) numbers of the moving-bed reactor and the reaction network has: (i) $ng=7$ gas phase components ($n_g^R = 6$) reactive components and ($n_g^i = 1$) inert component: N_2), (ii) ($n_s = n_s^R = 2$) solid phase reactive components (raw biomass $C_aH_bO_d$ and char biomass C), (iii) the $m=8$ reaction kinetics presented in Table 3, (iv) the corresponding reaction rate functions $(_{Rj})$ listed in Table 4, and (v) the stoichiometry-based reaction set shown in Table 5.

Table 3: Reaction kinetics of the biomass gasifier case example

Biomass pyrolysis
$C_aH_bO_d \xrightarrow{k_1} 0.54C + 0.12CO + 0.083CO_2 + 0.098H_2 + 0.273CH_4 + 0.064H_2O$
Char combustion
$C + O_2 \xrightarrow{k_2} CO_2$
Char reduction
$C + CO_2 \xrightarrow{k_3} 2CO$
$C + H_2O \xrightarrow{k_4} CO + H_2$
Water gas shift
$CO + H_2O \xrightarrow{k_5} CO_2 + H_2$
$CO_2 + H_2 \xrightarrow{k_6} CO + H_2O$
Gas oxidation
$2CO + O_2 \xrightarrow{k_7} 2CO_2$
$2H_2 + O_2 \xrightarrow{k_8} 2H_2O$

Table 4: Reaction rate functions

$R_j = \kappa_j(T) f_j(\mathbf{A})$		
R_j	$\kappa_j(T)$	$f_j(\mathbf{A})$
R1	k_1	A_1^x
R2	$\nu_p \left(\dfrac{k_2 k_m}{k_2 + k_m} \right)$	A_1^g
R3	$\nu_p \left(\dfrac{k_3 k_m}{k_3 + k_m} \right)$	A_4^g
R4	$\nu_p \left(\dfrac{k_4 k_m}{k_4 + k_m} \right)$	A_5^g
R5	k_5	$A_3^g A_5^g$
R6	k_6	$\bar{A}_2^g \bar{A}_4^g$
R7	k_7	$A_1^g A_3^g (A_5^g)^{1/2}$
R8	k_8	$A_1^g A_2^g$
$k_j = L_j e^{-E_j/UT}$		
Lj, Ej: Di Blasi (2000)		
$k_m = \dfrac{2.06 u_g}{\alpha_g} \mathrm{Re}^{-0.575} Sc^{-(2/3)}$		
$Sc = \dfrac{\mu_g}{\rho_g D_M}, \ \mathrm{Re} = \dfrac{d_p \rho_g u_g}{\mu_g}$		
$d_p = [(1 - \chi_{ash}) \chi_C + \chi_{ash}]^{1/3} d_{p0}$		
$X_C = \dfrac{u_s}{u_{s0}}, \ \nu_p = \dfrac{6(1 - \alpha_g)}{d_p}$		

Table 5: Reaction set of the biomass gasification case

Reaction (R_j)	$+s_{1,j}^s \Omega_1^s$	$+s_{2,j}^s \Omega_2^s$	$+s_{1,j}^g \Omega_1^g$	$+s_{2,j}^g \Omega_2^g$	$+s_{3,j}^g \Omega_3^g$	$+s_{4,j}^g \Omega_4^g$	$+s_{5,j}^g \Omega_5^g$	$+s_{6,j}^g \Omega_6^g$	$=0$
Biomass pyrolysis (R_1)	$-C_aH_bO_d$	$+0.54C$		$+0.098H_2$	$+0.119CO$	$+0.083CO_2$	$+0.273H_2O$	$+0.064CH_4$	$=0$
Char combustion (R_2)		$-C$	$-O_2$			$+CO_2$			$=0$
Char gasification 1 (R_3)		$-C$			$+2CO$	$-CO_2$			$=0$
Char gasification 2 (R_4)		$-C$		$+H_2$	$+CO$	$-CO_2$	$-H_2O$		$=0$
Water gas shift 1 (R_5)				$+H_2$	$-CO$	$+CO_2$	$-H_2O$		$=0$
Water gas shift 2 (R_6)				$-H_2$	$+CO$	$-CO_2$	$+H_2O$		$=0$
Gas combustion 1 (R_7)			$-O_2$		$-2CO$	$+2CO_2$			$=0$
Gas combustion 2 (R_8)			$-O_2$	$-2H_2$			$+2H_2O$		$=0$

The reaction network (Table 3, Table 4 and Table 5) has from two to four less reactions compared to the related previous studies (Rogel and Aguillon, 2006 and Di Blasi, 2000).

The overall gas molar concentration (Cg) and density (ρ_g) were determined with the ideal gas law, according to the expression

$$C_g = \frac{P_g}{U\,T}, \quad \rho_g = C_g W_g$$

(4)

where P_g, W_g are the pressure and the average molecular weight of the gas phase, respectively.

The experiment-based correlations for the transport properties $(\mu_g, \lambda_g^{eff}, \lambda_s^{eff})$ and the operating conditions were taken from Di Blasi (2000).

MODEL DEVELOPMENT

In this section, a reduced-order model design method for moving-bed gasification reactors is presented.

First, a reduced-order model of the reactor distributed system (2a), (2b), (2c), (2d), (2e), (2f), (2g) and (2h)is developed by combining assumptions, notions and tools from tubular chemical reactor engineering and reaction network theory: (i) the gas phase is in quasi-steady state regime (Baldea and Daoutidis, 2007 and Varma and Amundson, 1973), (ii) there is large interphase heat transfer rate (Caram and Fuentes, 1982 and Gobel et al., 2007), (iii) a suitable concentration linear coordinate change, determined by stoichiometry, takes the dynamic mass balance into a form with transport-reactive and transport parts (Aris, 1965 and Amundson and Arri, 1978), and (iv) the application of finite differences spatial discretization leads to a lumped model equivalent to a train of CSTRs with backflow (Coste et al., 1961 and Aris, 1961).

Finally, the number of CSTRs and their volumes of the reduced model constitute the design degrees of freedom that are chosen according to the modeling purpose and experimental uncertainty.

Mass Balances in Reactive–Nonreactive Form

This section is dedicated to the extension of stoichiometry-based notions from reaction network theory (Aris, 1965 and Feinberg, 1977) to the two-phase moving-bed reactor with two goals: (i) find a maximum set of independent reactions and (ii) to draw a model with minimum dependency on the nonlinear reaction rates that facilitates further stability analysis, control and observer design.

The components molar balances (2a) and (2c) of the moving-bed gasifier model (2a), (2b), (2c), (2d), (2e),(2f), (2g) and (2h) expressed in compact matrix-vector form are

$$\partial_{t'} \mathbf{A}_s = \Gamma_M^s \{\mathbf{A}_s\} + \mathbf{S}_s^T \mathbf{R}(\mathbf{A}, \mathbf{T}) \tag{5a}$$

$$\partial_{t'} \mathbf{A}_g = \Gamma_M^g \{\mathbf{A}_g\} + \mathbf{S}_g^T \mathbf{R}(\mathbf{A}, \mathbf{T}) \tag{5b}$$

$$\mathbf{S}_s = [s_{i,j}^s], \quad dim(\mathbf{S}_s) = (m, n_s^R) \tag{5c}$$

$$\mathbf{S}_g = [s_{i,j}^g], \quad dim(\mathbf{S}_g) = (m, n_g^R) \tag{5d}$$

where

$$R(\mathbf{A}, \mathbf{T}) = [R_1, \quad R_2, ..., R_m]^T$$

$\mathbf{S}_s^T$ (or $\mathbf{S}_s^T$) denotes the transpose of the solid (or gas) phase stoichiometric matrix.

In a more compact form equation (5a), (5b), (5c) and (5d) is written as follows:

$$\partial_t \mathbf{A} = \Gamma_M\{\mathbf{A}\} + \mathbf{S}^T \mathcal{R}(\mathbf{A}, \mathbf{T}), \qquad \Gamma_M\{\mathbf{A}\} = \begin{bmatrix} \Gamma_M^s\{\mathbf{A}_s\} \\ \Gamma_M^g\{\mathbf{A}_g\} \end{bmatrix}$$

(6a)

$$\mathbf{S}^T = \begin{bmatrix} \mathbf{S}_s^T \\ \mathbf{S}_g^T \end{bmatrix}, \qquad rank(\mathbf{S}) = N_r \le m, \quad n_g^R + n_s^R - N_r = \iota_r$$

(6b)

where S is the augmented stoichiometric matrix.

According to the reaction network theory (Aris, 1965 and Feinberg, 1977), the evaluation of the rank of the stoichiometric matrix S 6b) leads to a simplified model representation in the sense that: (i) only a set of Nr independent reactions must be considered and (ii) there are $n_s^R + n_g^R$ combinations of solid (5a) and gas (5b) mass balances to obtain a partition of Nr reactive and $_r$ non-reactive pseudo-species (linear combinations of species concentrations) balances.

Straightforward application of the reactive–non-reactive representation to the mass balances of moving-bed gasifier (5a) and (5b) would combine solid and gas dynamics in pseudo-species balances, preventing from reducing the order of the distributed system (2a), (2b), (2c), (2d), (2e), (2f), (2g) and (2h) through the enforcement of the QSS assumption in gas phase. Therefore, this model representation is performed for each phase in the present approach.

Independent Reaction Set

In principle, a set of Nr independent reactions (linear combinations of original reactions) can be obtained from the reduced row echelon form of matrix S by applying a Gauss–Jordan-elimination procedure (Aris, 1965 and Feinberg, 1977).

The present study takes advantage of the rank factorization based on reduced row echelon form of S(7a)to attain a new truncated stoichiometric matrix S and reaction rates vector N in terms of the set of independent reactions

$$\mathbf{S}^T = \mathcal{S}^T \, \mathcal{M} \tag{7a}$$

$$dim(\mathcal{S}) = (N_r, n_s^R + n_g^R), \qquad dim(\mathcal{M}) = (N_r, m)$$

$$\mathcal{S}^T = \begin{bmatrix} \mathcal{S}_s^T \\ \mathcal{S}_g^T \end{bmatrix}, \qquad dim(\mathcal{S}_\pi) = (N_r, n_\pi^R),$$

$$rank(\mathcal{S}_\pi) = N_\pi = \min(N_r, n_\pi^R) \quad \pi = s, g \tag{7b}$$

where S is formed by the Nr pivot columns of S and M is made by the Nr non-zero rows of the reduced row echelon form of S. Accordingly, the rate of generation vector

$$\mathcal{G}_\pi(\mathbf{A}, \mathbf{T}) = \mathcal{S}_\pi^T \mathcal{N}(\mathbf{A}, \mathbf{T}), \mathcal{N}(\mathbf{A}, \mathbf{T}) := \mathcal{M}\mathcal{R}(\mathbf{A}, \mathbf{T}) \tag{8}$$

can be written in terms of the new vector N with stoichiometrically independent reaction rates.

Then, the new stoichiometric matrices $(_{Ss'} _{Sg})$ with full rank (7b) are the point of departure of the following reactive–non-reactive representation of mass balances.

Reactive and Non-Reactive Components Based Representation

Following the stoichiometry-based reaction network theory (Aris, 1965 and Feinberg, 1977), if $n_s^R > N_s$ and $n_g^R > N_g$, there exists a suitable set of n_g^R linear combinations of the nRs solid species concentrations and nRg linear combinations of the nRg gas components concentrations, or equivalently, a coordinate change

$$\mathbf{C}_s := \begin{bmatrix} \mathbf{W}_s \\ \mathbf{E}_s \end{bmatrix} = \mathbf{P}_s \mathbf{A}_s, \quad \mathbf{C}_g := \begin{bmatrix} \mathbf{W}_g \\ \mathbf{E}_g \end{bmatrix} = \mathbf{P}_g \mathbf{A}_g \tag{9}$$

such that the molar balances (5a), (5b), (5c) and (5d) are taken into the reactive and non-reactive pseudo-species form

$$\partial_{t'} \mathbf{W}_s = \Gamma_M^s \{\mathbf{W}_s\}(\mathbf{C}, \mathbf{T}) + \mathbf{R}_s(\mathbf{C}, \mathbf{T})$$

(10a)

$$\partial_{t'} \mathbf{E}_s = \Gamma_M^s \{\mathbf{E}_s\}(\mathbf{C}, \mathbf{T})$$

(10b)

$$\partial_{t'} \mathbf{W}_g = \Gamma_M^g \{\mathbf{W}_g\}(\mathbf{C}, \mathbf{T}) + \mathbf{R}_g(\mathbf{C}, \mathbf{T})$$

(10c)

$$\partial_{t'} \mathbf{E}_g = \Gamma_M^g \{\mathbf{E}_g\}(\mathbf{C}, \mathbf{T})$$

(10d)

$$\mathbf{R}_s(\mathbf{C}, \mathbf{T}) = \mathbf{P}_s \mathcal{S}_s^T \mathcal{N}(\mathbf{A}_s, \mathbf{A}_g, \mathbf{T})$$

(10e)

$$\mathbf{R}_g(\mathbf{C}, \mathbf{T}) = \mathbf{P}_g \mathcal{S}_g^T \mathcal{N}(\mathbf{A}_s, \mathbf{A}_g, \mathbf{T})$$

(10f)

$$\mathbf{C} = [\mathbf{W}_s^T, \quad \mathbf{E}_s^T, \quad \mathbf{W}_g^T, \quad \mathbf{E}_g^T]^T$$
$$\mathbf{A}_s = \mathbf{P}_s^{-1} \begin{bmatrix} \mathbf{W}_s \\ \mathbf{E}_s \end{bmatrix}, \quad \mathbf{A}_g = \mathbf{P}_g^{-1} \begin{bmatrix} \mathbf{W}_g \\ \mathbf{E}_g \end{bmatrix}$$

(10g)

where

$$\mathbf{P}_\pi = \begin{bmatrix} \mathcal{T}_\pi & \mathbf{0}_{n_\pi^R \times n_\pi^l} \\ \mathbf{0}_{n_\pi^R \times n_\pi^l}^T & \mathbf{I}_{n_\pi^l} \end{bmatrix}, \quad \mathcal{T}_\pi = \begin{bmatrix} \mathcal{P}_\pi^R \\ \mathcal{P}_\pi^N \end{bmatrix}$$

(11a)

$$dim(\mathcal{P}_\pi^R) = (N_\pi, n_\pi^R), \quad dim(\mathcal{P}_\pi^N) = (n_\pi^R - N_\pi, n_\pi^R)$$

$$\mathbf{W}_\pi = [W_1^\pi, W_2^\pi, ..., W_{N_\pi}^\pi]^T \tag{11b}$$

$$\mathbf{E}_\pi = [\mathcal{E}_\pi^T, \mathcal{I}_\pi^T]^T, \quad \pi = g, s \tag{11c}$$

$$\mathcal{E}_\pi = [\mathcal{E}_1^\pi, \mathcal{E}_2^\pi, ..., \mathcal{E}_{n_\pi^R - N_\pi}^\pi]^T \tag{11d}$$

$$\mathcal{I}_\pi = [\mathcal{I}_1^\pi, \mathcal{I}_2^\pi, ..., \mathcal{I}_{n_\pi^I}^\pi]^T \tag{11e}$$

W_ϖ (or E_ϖ) contains the reactive (or non-reactive E_ϖ plus inert I_ϖ) concentration vector of phase ϖ, $I_{n_\pi^I}$ is the n_π^I identity matrix, $0_{n_\pi^R \times n_\pi^I}$ is the $n_\pi^R \times n_\pi^I$ zero matrix, p_π^R (or p_π^R) is the coordinate change matrix of the reactive key (or reactive) pseudo-components concentrations.

Matrices p_π^R and p_π^R, that determine the key-non-reactive concentration partition (11a), can be found by inspection (Aris, 1965 and Amundson and Arri, 1978) combining molar balance equations (2a) and (2c) to eliminate or reduce the nonlinear reaction term dependence of the maximum number of balances. However, the inspection method is not suitable if the reaction network is relatively complex.

A diversity of linear algebra tools can be applied to find the matrix P_ϖ based on range and null space decomposition of S_π^T. In view of the uncertainty on some of the stoichiometric coefficients, here a robustness-oriented coordinate change, based on singular value decomposition (SVD) will be employed. For this aim, the SVD of the truncated stoichiometric matrix S_π^T is written as

$$S_\pi^T = \mathbf{U}_\pi \Sigma_\pi \mathbf{V}_\pi^T$$

(12)

where $\mathbf{U}_\varpi$ (or $\mathbf{V}_\varpi$) is a $n_\pi^R \times n_\pi^R$ (or $_{N\varpi} \times _{N\varpi}$) unitary matrix with the left (or right) singular vectors of S_π^T as columns, $_\varpi$ is a diagonal matrix with the singular values of . Thus, the matrices p_π^R and p_π^R determined by means of the SVD method are

$$T_\pi = \begin{bmatrix} \mathcal{P}_\pi^R \\ \mathcal{P}_\pi^N \end{bmatrix} = \begin{bmatrix} \mathcal{S}_\pi^+ \\ \mathbf{L}_\pi \end{bmatrix}$$

(13)

where $\mathbf{L}_\pi^T$ is made by the last $_\varpi$ columns of $_{U\varpi}$ and $\mathcal{S}_\pi^+$ is obtained according to the following expression:

$$\mathcal{S}_\pi^+ = \mathbf{V}_\pi \Sigma_\pi^+ \mathbf{U}_\pi^T$$

(14)

where Σ_π^+ the pseudo-inverse of $_\varpi$ (obtained by taking the reciprocal of every non-zero diagonal entry, leaving zeros in place, and transposing the matrix).

In particular, when $n_\pi^R \leq N_\pi$, the coordinate change (9) is not needed in the ϖ-phase and the corresponding molar balances of (10a), (10c), (10d), (10e), (10f) and (10g) only have a reactive part because we have that

$$\mathbf{C}_\pi = \mathbf{W}_\pi = \mathbf{A}_\pi, \quad \mathbf{P}_\pi = \mathbf{I}_{n_\pi}$$

(15a)

$$\mathbf{R}_\pi(\mathbf{C}, \mathbf{T}) = \mathcal{S}_\pi \mathcal{N}(\mathbf{A}, \mathbf{T})$$

(15b)

Biomass Gasifier Stoichiometric Considerations

The stoichiometric matrix for the biomass gasifier case example is built according to Table

$$
\mathbf{S}^{T} =
\begin{bmatrix}
-1 & 0 & 0 & 0 & 0 & 0 & 0 & 0 \\
0.54 & -1 & -1 & -1 & 0 & 0 & 0 & 0 \\
0 & -1 & 0 & 0 & 0 & 0 & -1 & -1 \\
0.098 & 0 & 0 & 1 & 1 & -1 & 0 & -2 \\
0.119 & 0 & 2 & 1 & -1 & 1 & -2 & 0 \\
0.083 & 1 & -1 & 0 & 1 & -1 & 2 & 0 \\
0.273 & 0 & 0 & -1 & -1 & 1 & 0 & 2 \\
0.064 & 0 & 0 & 0 & 0 & 0 & 0 & 0
\end{bmatrix}
\tag{16}
$$

where $\mathbf{S}_s^T$ is made by the first $n_s^R = 2$ rows of $\mathbf{S}^T$ and $\mathbf{S}_g^T$ by the last $n_g^R = 6$ rows of $\mathbf{S}^T$.

It is found that there are $Nr=4$ independent reactions (or reactive concentrations) from rank test application(17) to the case study stoichiometric matrix (16)

$$
dim(\mathbf{S}) = (8, 8), \quad N_r = rank(\mathbf{S}) = 4, \quad t_r = 4
\tag{17}
$$

In agreement with the reduced row echelon form of $\mathbf{S}^T$, the new truncated stoichiometric matrix S for the case example is formed by the set of the first $Nr=4$ reactions $(_{R1} -_{R4})$ from the original matrix S(16)

$$S^T = \begin{bmatrix} S_s^T \\ S_g^T \end{bmatrix}, \quad dim(S) = (8,4), \quad rank(S) = 4 \tag{18}$$

where

$$S_s^T = \begin{bmatrix} -1 & 0 & 0 & 0 \\ 0.54 & -1 & -1 & -1 \end{bmatrix}, \quad rank(S_s) = 2$$

$$S_g^T = \begin{bmatrix} 0 & -1 & 0 & 0 \\ 0.098 & 0 & 0 & 1 \\ 0.119 & 0 & 2 & 1 \\ 0.083 & 1 & -1 & 0 \\ 0.273 & 0 & 0 & -1 \\ 0.064 & 0 & 0 & 0 \end{bmatrix}, \quad rank(S_g) = 4$$

According to this result there are four redundant reactions in the biomass gasification network (Table 5), meaning that the last four reactions (water gas shift and gas phase combustions) are dependent.

In the present study, the independent reaction set is provided by the rank factorization of matrix S^T(7a)based on its reduced row echelon form without zero rows

$$M = \begin{bmatrix} 1 & 0 & 0 & 0 & 0 & 0 & 0 & 0 \\ 0 & 1 & 0 & 0 & 0 & 0 & 1 & 1 \\ 0 & 0 & 1 & 0 & -1 & 1 & -1 & 1 \\ 0 & 0 & 0 & 1 & 1 & -1 & 0 & -2 \end{bmatrix} \tag{19}$$

The corresponding $Nr=4$ entry vector N with the independent reaction rates is given by equation (20), in the understanding that such representation is intended for robustness purposes and not for chemical interpretation

$$\mathcal{N}(\mathbf{A}, T) = \begin{bmatrix} \mathcal{R}_1 \\ \mathcal{R}_2 + \mathcal{R}_7 + \mathcal{R}_8 \\ \mathcal{R}_3 - \mathcal{R}_5 + \mathcal{R}_6 - \mathcal{R}_7 + \mathcal{R}_8 \\ \mathcal{R}_4 + \mathcal{R}_5 - \mathcal{R}_6 - 2\mathcal{R}_8 \end{bmatrix}$$

(20)

Consequently, the first independent reaction is the pyrolisis and the last three independent reaction are linear combinations of the remaining reactions in Table 3.

Given $n_g^R > N_g$, a partition of the $_g$=2 non-reactive components and the $_{Ng}$=4 reactive pseudo-species is obtained for the gas phase concentrations by means of the concentration linear coordinate change(11a) and (11c), which for the case example yields

$$\mathbf{P}_g = \begin{bmatrix} \mathcal{T}_g & \mathbf{0}_{6\times1} \\ \mathbf{0}_{1\times6} & 1 \end{bmatrix}, \quad \mathcal{T}_g = \begin{bmatrix} \mathcal{P}_g^R \\ \mathcal{P}_g^N \end{bmatrix}$$

(21)

where $\mathcal{P}_g^R$ is calculated from the SVD based pseudoinverse of $\mathcal{S}_g^T$(14)

$$\begin{bmatrix} 0.9 & 1.9 & 0.45 & 0.9 & 2.3 & 0.74 \\ -0.65 & -0.25 & 0.18 & 0.35 & -0.08 & -0.06 \\ -0.22 & -0.35 & 0.39 & -0.22 & 0.04 & -0.05 \\ 0.17 & 0.6 & 0.08 & 0.17 & -0.31 & 0.05 \end{bmatrix}$$

and $\mathbf{P}_g^N$ is made by the last two rows of $\mathbf{U}_g^T$ in the SVD of $\mathbf{S}_g^T$ (12)

$$\begin{bmatrix} -0.58 & 0.45 & -0.29 & -0.58 & 0.16 & -0.09 \\ -0.11 & -0.09 & -0.06 & -0.11 & -0.14 & 0.97 \end{bmatrix}$$

Since $N_s = n_s^R$ (15a) for the case example there is no solid species coordinate change, and consequently, we have that

$$\mathbf{C}_s = \mathbf{W}_s = \mathbf{A}_s, \quad \mathbf{A}_s = [A_1^s, \; A_2^s]^T$$

(22a)

$$\mathbf{R}_s(\mathbf{C}, \mathbf{T}) = \mathcal{G}_s = \mathcal{S}_s^T \mathcal{N}(\mathbf{A}, \mathbf{T})$$

(22b)

Distributed Model

In the present section, the stoichiometry-based considerations of Section 3.1, model reduction and simplifying assumptions are applied to the dimensionless form of moving-bed gasifier model (2a), (2b),(2c), (2d), (2e), (2f), (2g) and (2h).

Since the solid density is approximately 300 times the one of the gas (implying that $u_{us} \ll u_{ug}$) the QSS assumption for the gas phase holds (Gobel et al., 2007 and Baldea and Daoutidis, 2007). Additionally, large interphase heat transfer rate ($T_s = T_g = T$) is regarded (Gobel et al., 2007 and Caram and Fuentes, 1982). From this assumptions in conjunction with the preceding stoichiometric considerations the description of the moving-bed reactor dynamics (2a), (2b), (2c), (2d), (2e), (2f), (2g) and (2h), in terms of dimensionless variables (25a) and numbers (25b), becomes the set of PDEs

$$\partial_t \mathbf{W}_s = \tau_s^C \{\mathbf{W}_s\}(\mathbf{C}, \eta) + \mathbf{r}_s(\mathbf{C}, \eta)$$

(23a)

$$\partial_t \mathbf{e}_s = \tau_s^C \{\mathbf{e}_s\}(\mathbf{C}, \eta)$$

(23b)

$$\partial_t \eta = \tau_H^C\{\eta\}(\mathbf{c},\eta) + \frac{1}{Pe_H}\tau_H^D\{\eta\}(\mathbf{c},\eta) + Q_T(\mathbf{c},\eta)$$

(23c)

$$0 = \tau_g^C\{\mathbf{w}_g\}(\mathbf{c},\eta) + \frac{1}{Pe_M}\tau_g^D\{\mathbf{w}_g\}(\mathbf{c},\eta) + \mathbf{r}_g(\mathbf{c},\eta)$$

(23d)

$$0 = \tau_g^C\{\mathbf{e}_g\}(\mathbf{c},\eta) + \frac{1}{Pe_M}\tau_g^D\{\mathbf{e}_g\}(\mathbf{c},\eta)$$

(23e)

$$\mathbf{c} = [\mathbf{w}_s^T, \mathbf{e}_s^T, \mathbf{w}_g^T, \mathbf{e}_g^T]^T$$

$$\mathbf{a}_\pi = \frac{\mathbf{A}_\pi}{C_{\pi 0}} = \mathbf{P}_\pi^{-1}\begin{bmatrix}\mathbf{w}_\pi \\ \mathbf{e}_\pi\end{bmatrix}, \quad T = \eta T_{n0}, \quad \pi = g, s$$

(23f)

Boundary and initial conditions

$$z = 0 : \beta_g\{\mathbf{c}_g\} = 0, \quad \beta_s\{\mathbf{c}_s\} = 0, \quad \beta_H\{\eta\} = 0$$

(23g)

$$z = 1 : \partial_z \mathbf{c}_g = 0, \quad \partial_z \mathbf{c}_s = 0, \quad \partial_z \eta = 0$$

(23h)

$$t = 0 : \mathbf{c}_s(z, 0) = \frac{\mathbf{P}_s \mathbf{A}_{sic}}{C_{s0}}, \quad \eta(z, 0) = \frac{T_{ic}}{T_{n0}}$$

$$(23i)$$

and source terms

$$\mathbf{r}_s(\mathbf{c}, \eta) = K_M \mathcal{M} \, \mathbf{Da} \, \mathbf{r}(\mathbf{c}, \eta)$$

$$(24a)$$

$$\mathcal{Q}_T(\mathbf{c}, \eta) = -St_w(\eta)[\eta - \eta_w] - K_M \mathbf{B}_r^T \mathbf{Da} \, \mathbf{r}(\mathbf{c}, \eta)$$

$$(24b)$$

$$\mathbf{r}_g(\mathbf{c}, \eta) = \mathcal{M} \, \mathbf{Da} \, \mathbf{r}(\mathbf{c}, \eta)$$

$$(24c)$$

where

$$\mathbf{r} = [r_1, \; r_2, \; \ldots, \; r_m]^T, \quad r_j(\mathbf{c}, \eta) = \frac{\mathcal{R}_j(\mathbf{C}, T)}{\mathcal{R}_j(\mathbf{C}_{g0}, T_{n0})}$$

$$\mathbf{C}_{g0} = [C_{g0}, C_{g0}, \ldots, C_{g0}]^T$$

$$(24d)$$

The dimensionless variables and numbers are

$$\mathbf{w}_g = \frac{\mathbf{W}_g}{C_{g0}}, \quad \mathbf{w}_s = \frac{\mathbf{W}_s}{C_{s0}}, \quad \mathbf{e}_g = \frac{\mathbf{E}_g}{C_{g0}}, \quad \mathbf{e}_s = \frac{\mathbf{E}_s}{C_{s0}}, \quad z = \frac{Z}{L}$$

$$v_g = \frac{u_g}{u_{g0}}, \quad v_s^n = \frac{u_s}{u_{s0}}, \quad \eta = \frac{T}{T_{n0}}, \quad \eta_w = \frac{T_w}{T_{n0}}, \quad t = \frac{u_{s0}t'}{L}$$

$$K_M = \frac{C_{g0}u_{g0}}{C_{s0}u_{s0}}, \quad D_M = \frac{\mathcal{D}_M}{\mathcal{D}_{M0}}, \quad D_H = \frac{\mathcal{D}_H}{\mathcal{D}_{H0}}.$$

$$\mathcal{D}_H = \frac{\lambda_s^{eff} + \lambda_g^{eff}}{C_s C_p^s}, \quad \mathcal{D}_{H0} = \frac{\lambda_0}{C_s C_p^s}$$

$$(25a)$$

$$Pe_M = \frac{L u_{g0}}{\mathcal{D}_{M0}}, \quad Pe_H = \frac{L u_{s0}}{\mathcal{D}_{H0}}, \quad St_w = \frac{L h_w A_v}{C_s C_p^s u_{s0}}$$

$$Da_j = \frac{L R_j(\mathbf{C}_{g0}, T_{n0})}{u_{g0} C_{g0}}, \quad B_j = \frac{\Delta H_j^r}{C_p^s T_{n0}}$$

$$(25b)$$

with suitable reference values in the denominators of (25a). Here, DH is the dimensionless total heat diffusion coefficient, PeM (or PeH) is the gas mass (or heat) Peclet number, Stw is the wall Stanton number, Da is a diagonal matrix with m Damkholer numbers (Daj), $_{Br}$ is the vector of the m adiabatic temperature rises (Bj), and r is the reaction rate vector.

The dimensionless convection transport operators of mass (τ_g^C, τ_s^C)and total heat τ_H^C, the diffusion transport operators of mass τ_g^D and heat τ_H^D, the dimensionless total heat convection coefficient $_{vH}$ and the boundary operators (β_g, β_s and β_H) are defined in Table 6.

Table 6: Operators of the dimensionless form of the reduced model (23a), (23b), (23c), (23d), (23e), (23f), (23g), (23h) and (23i)

Boundary operators
$\beta_g\{\mathbf{X}\} := \frac{\mathcal{D}_M}{Pe_M}\,\partial_z \mathbf{X} - V_g[\mathbf{X} - \mathbf{X}_e]$
$_{\beta s}\{\mathbf{x}\} := \mathbf{x} - _{xe}$
$\beta_s\{\mathbf{X}\} := \mathbf{X} - \mathbf{X}_e$
Transport operators
$\tau_g^C\{\mathbf{X}\}(\mathbf{C}, \eta) := -\partial_z(V_g(\mathbf{C}, \eta)\,\mathbf{X})$
$\tau_g^D\{\mathbf{X}\}(\mathbf{C}, \eta) := \partial_z(\mathcal{D}_M(\mathbf{C}, \eta)\,\partial_z \mathbf{X})$
$\tau_s^C\{\mathbf{X}\}(\mathbf{C}, \eta) := -\partial_z(V_s(\mathbf{C}, \eta)\,\mathbf{X})$

$$r_H^C\{\eta\}(\mathbf{c},\eta):=-\partial_z(v_H(\mathbf{c},\eta)\,\eta)$$

$$r_H^D\{\eta\}(\mathbf{c},\eta)=\partial_z(D_H(\mathbf{c},\eta)\,\partial_z\eta)$$

$$v_H = v_s + K_M\left[\frac{c_p^g c_g}{c_p^s c_s}\right]v_g$$

The vector $_{cg,e}$ (or $_{cs,e}$) contains the dimensionless gas (or solid) feed concentrations, respectively, η_e is the dimensionless average feed temperature of the gas and solid phases. $_{Asic}$ is the vector with the initial solid-phase concentrations, *Tic* is the initial temperature profile, and *hw* is the wall heat transfer coefficient, and *DR* is the reactor diameter.

The preceding distributed system model (23a), (23b), (23c), (23d), (23e), (23f), (23g), (23h) and (23i) is a reduced-order version of previous syngas tubular reactor models (Di Blasi, 2000, DiBlasi, 2004 and Grieco and Baldi, 2011) of the form (2a), (2b), (2c), (2d), (2e), (2f), (2g) and (2h), with less dynamic equations $_{ns}$+1 PDEs and *ng* ODEs.

Lumped Model

In this section the reactor is represented, via spatial discretization plus interpolation, as a train of *N* adjustable-volume CSTRs, which is the point of departure of the proposed model design method presented in Section 3.4.

To obtain the simplest possible lumped representation of the moving-bed gasifier model (23a), (23b),(23c), (23d), (23e), (23f), (23g), (23h) and (23i) such that the distributed nature of the reactor is retained with the weakest coupling of ODEs, the spatial discretization of reactor must be performed using any numerical method capable of preserving the dispersion–convection transport mechanism, that is, the dynamics of variables at *k* -th node of spatial mesh must only depend on variables at neighbor nodes (k+1 and k−1) and themselves.

For the sake of simplicity, the spatial discretization of the distributed reactor model (23a), (23b), (23c),(23d), (23e), (23f), (23g), (23h) and (23i) is performed here with first order finite-differences over an adjustable mesh of size Nz. The resulting dynamic model consists of N lumps (internal nodes of mesh:$_{Nz}$=N+2) with M=N×($_{ns}$+1) ODEs coupled with Nv=N×($_{ng}$) algebraic equations (AEs).

The solution of the Nv algebraic equations for the quasi-static (**y**) pseudo-species concentrations in terms of the dynamic ones and the temperature (**x**) followed by substitution into the M ODEs, yields the lumped model in staged system form

$$\dot{\mathbf{x}}_1 = \mathbf{f}_1(\mathbf{x}_1, \mathbf{x}_2, \mathbf{u})$$

(26a)

$$\dot{\mathbf{x}}_k = \mathbf{f}_k(\mathbf{x}_{k-1}, \mathbf{x}_k, \mathbf{x}_{k+1}), \quad k = 2, ..., N-1$$

(26b)

$$\dot{\mathbf{x}}_N = \mathbf{f}_N(\mathbf{x}_{N-1}, \mathbf{x}_N)$$

(26c)

$$\mathbf{y}_j = \mathbf{h}_j(\mathbf{x}_{j-1}, \mathbf{x}_j), \quad j = 1, ..., N$$

(26d)

where

$$\mathbf{x}_j = \begin{bmatrix} \mathbf{w}_{s,j} \\ \mathbf{e}_{s,j} \\ \eta_j \end{bmatrix}, \quad \mathbf{y}_j = \begin{bmatrix} \mathbf{w}_{g,j} \\ \mathbf{e}_{g,j} \end{bmatrix}, \quad \mathbf{u} = \begin{bmatrix} \mathbf{x}_e \\ \mathbf{y}_e \\ \mathbf{q}_e \end{bmatrix}, \quad j = 1, ..., N, e$$

(27)

$x_1, \ldots, x_N$ are the dynamic states, **u** is the exogenous input with the feed flows vector q_e, the feed values of the dimensionless gas pseudo-species concentrations (y_e), solid pseudo-species concentrations and temperature (x_e).

The dynamic variables vector x_k contains the dimensionless solid phase pseudo-species concentrations ($w_{s,j}$, $e_{s,j}$) and temperature ($_j$) values at the k-th node of spatial mesh, while the static variables vector y_k includes the dimensionless gas phase pseudo-species concentrations ($w_{g,j}$, $e_{g,j}$) values at the k-th node of mesh.

N-CSTR Model

Following the classical tubular reactor lumping of Coste et al. (1961), Aris (1961), and Deans and Lapidus (1960), and the recent study of Najera (2012) for an exothermic single-component tubular reactor, which exhibits a similar behavior (multiplicity of steady-states, bifurcation phenomena) to the one displayed by the moving-bed gasifier, the rearrangement of the resulting set of differential–algebraic equations obtained from the finite differences spatial discretization of distributed model (23a), (23b), (23c), (23d), (23e), (23f),(23g), (23h) and (23i) yields the N-CSTR model (see equation set (38a), (38b), (38c), (38d), (38e), (38f),(38g), (38h), (38i) and (38j) of Appendix) with volumes $v_1, v_2, \ldots, v_N$ as depicted in Fig. 2.

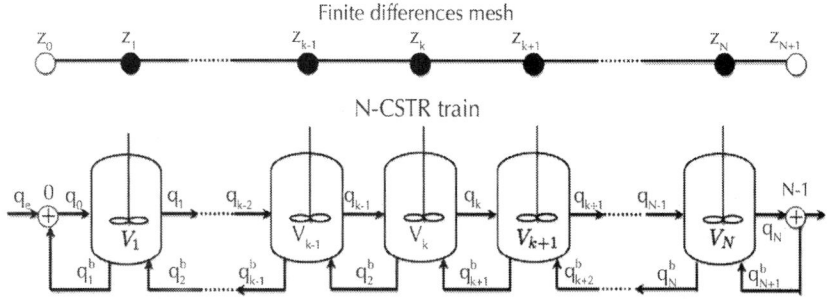

Figure 2: Analogy between spatial discretization mesh and a train of CSTRs with back-mixing flows.

The convective velocities (v_ϖ and v_H) and the diffusion coefficients (DM and DH) of the tubular reactor(23a), (23b), (23c), (23d), (23e), (23f), (23g), (23h) and (23i) are related with the forward (q_k) and back-mixing q_k^b flow vectors of the CSTR set (38a), (38b), (38c), (38d), (38e), (38f), (38g), (38h), (38i) and (38j)as follows:

$$\mathbf{q}_k = \begin{bmatrix} \mathbf{q}_k^{Ms} \\ \mathbf{q}_k^{H} \\ \mathbf{q}_k^{Mg} \end{bmatrix} = \begin{bmatrix} v_{s,k}A_R \\ v_{H,k}A_R + \dfrac{D_{H,k}A_R}{Pe_H \Delta z_{k+1}} \\ v_{g,k}A_R + \dfrac{D_{M,k}A_R}{Pe_M \Delta z_{k+1}} \end{bmatrix}$$

$$(28a)$$

$$\mathbf{q}_k^b = \begin{bmatrix} \mathbf{q}_k^{b.Ms} \\ \mathbf{q}_k^{b.H} \\ \mathbf{q}_k^{b.Mg} \end{bmatrix} = \begin{bmatrix} 0 \\ \dfrac{D_{H,k}A_R}{Pe_H \Delta z_{k+1}} \\ \dfrac{D_{M,k}A_R}{Pe_M \Delta z_{k+1}} \end{bmatrix}$$

$$(28b)$$

The exit boundary conditions (at $z=1$) yield the trivial algebraic expression (29b), and the inlet ones (at $z=0$) yield a set of algebraic equations that amount to mass and energy flow mixers (29a)

$$\mathbf{q}_0 \phi_0 = \mathbf{q}_e \phi_e + \mathbf{q}_1^b \phi_1 \quad \text{at } z = 0$$

$$(29a)$$

$$\phi_{N+1} = \phi_N \quad \text{at } z = 1$$

$$(29b)$$

Where

$$\mathbf{q}_e = \begin{bmatrix} \mathbf{q}_e^{Ms} \\ \mathbf{q}_e^{H} \\ \mathbf{q}_e^{Mg} \end{bmatrix} = \begin{bmatrix} v_{s,e}A_R \\ v_{H,e}A_R \\ v_{g,e}A_R \end{bmatrix}, \quad \phi_k = \begin{bmatrix} \mathbf{w}_{s,k} \\ \mathbf{e}_{s,k} \\ \eta_k \\ \mathbf{w}_{g,k} \\ \mathbf{e}_{g,k} \end{bmatrix}, \quad k = 0, \ldots, N+1, e$$

(30)

$v_{g,e}$ (or $v_{s,e}$) is the feed flow velocity of gas (or solid) phase, and ϕ_k is the vector with concentrations and the temperature at the k-th node.

The application of the proposed approach to the tubular reactor case example (Section 2.3) yields a N-CSTR model of the form (38a), (38b), (38c), (38d), (38e), (38f), (38g), (38h), (38i) and (38j), with convective flow velocities which are obtained from the expressions (Di Blasi, 2000)

$$v_{g,k}c_{g,k} = v_{g,k-1}c_{g,k-1} + \sum_{i=1}^{n_g} \sum_{j=1}^{NR} \nu_{i,j} Da_{j,k} r_{j,k}$$

(31a)

$$v_{s,k} = v_{s,k-1} - \frac{(Da_{G1,k}r_{G1,k} + Da_{G2,k}r_{G2,k} + Da_{C5,k}r_{C5,k})}{c_{C,pyr}}$$

(31b)

according to the following rationale.

Most of the previous models (Amundson and Arri, 1978, Di Blasi, 2000 and Gobel et al., 2007) of moving-bed gasifiers do not incorporate the momentum balance because of the fast hydrodynamic characteristic time scale in relation to the energy and mass time scales. Accordingly, in our case study the convective velocity of gas phase (32a) and char (32b) satisfy the continuity equations

$$\partial_z(C_g V_g) = \sum_{i=1}^{n_g} \sum_{j=1}^{m} s_{i,j}^g Da_j r_j$$

(32a)

$$a_{C,pyr} \partial_z V_s = -(Da_{G1} r_{G1} + Da_{G2} r_{G2} + Da_{C5} r_{C5})$$

(32b)

where $a_{C,pyr}$ is the dimensionless char concentration or the solid concentration after pyrolysis.

These expressions state that the pyrolysis reactions affect only the density of the solid, and that the char combustion and gasification produce structural changes modeled as velocity variations through the char continuity equation (32b).

The discretization of the continuity equations (32a) and (32b) yields (31a) and (31b).

Profile Approximation

The N-CSTRs approximation (38a), (38b), (38c), (38d), (38e), (38f), (38g), (38h), (38i) and (38j) of the distributed tubular reactor (23a), (23b), (23c), (23d), (23e), (23f), (23g), (23h) and (23i) yields the evolution of the variables in the Nz vectors $_k$ (30), with the $_{ns}+_{ng}$ concentrations and the temperature at the k-th node of the spatial mesh. The approximation of the concentration and temperature spatial profiles is obtained with standard spline interpolation, according to the following scheme (de Boor, 2001).

The $_k$ vectors (30) are grouped into the matrix

$$\Phi(t) = [\phi_0(t), \phi_1(t), \ldots, \phi_{N+1}(t)]$$

(33)

where the j-th row of matrix Φ, denoted by $\Phi_{j'}$, contains the sequence of the Nz nodal values of the j-th model variable.

The time evolution of the $ns+ng+1$ concentration and temperature profiles is given by

$$\Phi_j(t,z) = \sum_{r=1}^{N_z} a_r^j(t)B_{2,r}(z), \quad j=1,\ldots,n_s+n_g+1$$

(34)

where $B_{2,r}(z)$ is the r-th quadratic spline basis function

$$
\begin{cases}
\dfrac{(z-x_r)^2}{(x_{r+2}-x_r)(x_{r+1}-x_r)}, & x_r < z \leq x_{r+1} \\[2mm]
\dfrac{(z-x_r)(x_{r+2}-z)}{(x_{r+2}-x_r)(x_{r+2}-x_{r+1})} + \dfrac{(x_{r+3}-z)(z-x_{r+1})}{(x_{r+3}-x_{r+1})(x_{r+2}-x_{r+1})}, & x_{r+1} < z \leq x_{r+2} \\[2mm]
\dfrac{(x_{r+3}-z)^2}{(x_{r+3}-x_{r+2})(x_{r+3}-x_{r+1})}, & x_{r+2} < z \leq x_{r+3} \\[2mm]
0 & \text{otherwise}
\end{cases}
$$

(35)

over the knot sequence: $x_1 \leq x_2 \leq \cdots \leq x_{Nz+2} \leq x_{Nz+3}$ defined in terms of nodes of spatial mesh $(z_0=0 < z_1 < \cdots < z_N < z_{N+1}=1)$ as follows:

$$
x_r = \begin{cases}
z_0, & 1 \leq r \leq 3 \\[2mm]
\dfrac{z_{r-3}+z_{r-2}}{2}, & 4 \leq r \leq N_z \\[2mm]
z_{N+1}, & N_z+1 \leq r \leq N_z+3
\end{cases}
$$

(36)

The coefficient set $a_r^j(t)$ is the solution to the interpolation algebraic equations

$$\sum_{r=1}^{N_z} a_r^j(t)B_{2,r}(z_k) = \Phi_{j,k}(t), \quad k=1,\ldots,N_z$$

(37)

and $\Phi_{j,k}(t)$ is the value of the j-th variable at the k-th node.

Model Design Problem

In this section, the problem of designing a dynamic model with a smallest possible dimension (number of ODEs) is addressed in the light of a specific modeling purpose and the experimental uncertainty.

The number of tanks (N) and their volumes $(v_1, v_2, ..., v_N)$ are chosen with a two-step tuning procedure, according to the prescribed modeling objective of the N-CSTR model.

First, the volumes are set equal $(v_1 = v_2 = \cdots = v_N = V)$ and the number of tanks N is increased until the output variables behavior is described with a certain degree of accuracy, related to the modeling requirements. Then, on the basis of the resulting profile shapes, the number (N) of tanks is decreased and their volumes $(v_1, v_2, ..., v_N)$ are adjusted to meet the modeling objective with the smallest number of tanks.

The N-tank model with equal volumes constitutes a first reduced order model that facilitates the realization of the second design step because gives an initial estimate of the smallest number of tanks (N) and the corresponding volumes $(v_1, v_2, ..., v_N)$ needed to achieve the modeling demands in terms of accuracy. Consequently, if the first step of the design procedure is omitted, then the design of N and $v_1, v_2, ..., v_N$ based on the spatial profile shapes obtained from a more detailed model previously developed in the literature, would be a feasible but more complex modeling task.

Table 7: Comparison of model lumping procedures

Characteristic	(Di Blasi, 2000 and DiBlasi, 2004)	(Gobel et al., 2007)	(Rogel and Aguillon, 2006)	Present study
# of chemical components	$ns=3$, $ng=8$	$ns=2$, $ng=6$	$ns=3$, $ng=7$	$ns=3$, $ng=7$, $Ng=5$
# of chemical reactions	$m=12$, $msg=5$	$m=4$, $msg=2$	$m=10$, $msg=4$	$m=8$, $msg=4$
# of PDEs	14 PDEs	3 PDEs + 1 ODE	18 PDEs	3 PDEs + 9 ODEs
Gas QSS assumption	No	Yes	No	Yes
Lumping (N) # of internal nodes	Finite differences($N\approx250-500$)	Finite volume (Nnot available)	CFD Finite volume 2D mesh (5×20) ($N\approx100$)	3-tank model 9 ODEs + 27 AEs

NUMERICAL SIMULATIONS

According to reported simulation and experimental results (Di Blasi, 2000 and Barrio et al., 2001), downdraft gasifiers exhibit two operation modes (pyrolysis and combustion zones with different locations): top stabilized and grate-stabilized. The proposed modeling design approach was applied with the following objective: approximation of the top-stabilized steady state, which is the most effective mode of operation (Di Blasi, 2000), with the smallest possible number of ODEs and description capability similar to the one of the previous studies with PDE numerical approximation methods. As a reference, the first three columns of Table 8 summarize the results of the previous simulation (Di Blasi, 2000 and Rogel and Aguillon, 2006) and experimental (Manurung and Beenackers, 1993) studies.

Table 8: Comparison between model predictions versus experimental data (Manurung and Beenackers, 1993)

Variable	Exp. results[a]	Model with FD[b]	2D CFD model[c]	3-tank model (present study)
$\%_{X_{CO,N+1}}$	17–18	18.5–20.3	20–28	18.4
$\%_{X_{H2,N+1}}$	11–13.5	9.8–16.8	5.56–10	12.8
$\%_{X_{CO2,N+1}}$	11–13	9.4–15.3	9.78–10.75	14
$\%_{X_{CH4,N+1}}$	–	2.4–4.5	4–7	3.5
$\%_{X_{H2O,N+1}}$	–	8.8	10.5–11	8
$\%_{X_{N2,N+1}}$	45–55	43–60	46.9–47.2	43.5
$T_{g,N+1}$ (K)	837	810–942	741	900
$\dot{m}_g$ (kg/h)	–	44	–	43.25
LHV/HHV [d]	–	5.07/5.48	6.59/7.19	4.61/5.0

[a](Manurung and Beenackers, 1993) (experimental).

[b](Di Blasi, 2000) (simulation).

[c](Rogel, 2007) (simulation).

[d](MJ/Nm^3) normal condition: Tg=293.15 K, P=101.325 kPa.

The proposed N-CSTR model (38a), (38b), (38c), (38d), (38e), (38f), (38g), (38h), (38i) and (38j), with adjustable number of tanks and volumes, was set with the densities and transport parameters recalled from Di Blasi (2000), and initial conditions close to the top-stabilized operation mode.

Statics

In the first model design step, N=12 equal-volume tanks were needed to model the tubular reactor behavior with an accuracy similar to the one of the previous studies, including the location of the fast reaction rates based on the spatial profile shapes. Then, in the second design step, it was found that the modeling objective was met with N=3 tanks of different volumes selected on the basis of fast reaction rate locations (see Fig. 4): two small tanks ($_{V1}$=4/50$_{VR}$ and $_{V2}$=6/50$_{VR}$) to describe the steep profiles of the combustion and pyrolysis zones near the top and 1 large CSTR ($_{V3}$=4/5$_{VR}$) to describe the smooth profiles in the reduction zone.

Alternatively, the selection of the number of tanks and their volumes could be made according to the fast reaction rate locations derived from a more detailed model previously developed (Di Blasi, 2000). Consequently, the first step of the design procedure (i.e. the N=12 equal-volume tanks model) could be omitted and the modeling task would become more complicated due to the lack of an initial estimate of the size (i.e. N) for the reduced order model. The steady-state behavior of the 3 different-volume and 12 equal-volume tank models is presented inFig. 3 and Fig. 4, including (i) concentration, temperature and solid apparent density profiles with 3 and 12 tanks (Fig. 3) and (ii) convective gas and solid velocities as well as reaction rate profiles with the 3 tanks (Fig. 4). In Table 8 the results are compared with the ones of the previous studies with PDE solvers, showing that the 3-tank model matches (within parameter and experimental error) the behavior of the PDE models.

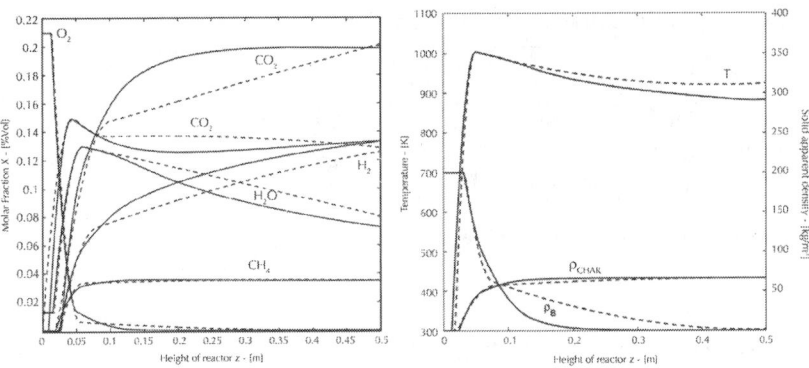

Figure 3: Steady state gas concentration, biomass and char densities, and temperature profiles with three different-volume (discontinuous plot) and 12 equal-volume (continuous plot) tank models.

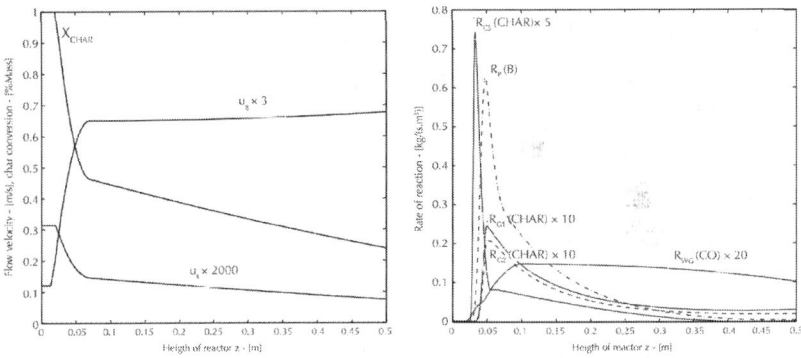

Figure 4: Steady state solid and gas convection velocities, char yield and reaction rate profiles with the 3-tank model.

Dynamics

To assess the 3 and 12-tank dynamic model behavior, a transient was induced by setting all the tanks at an initial conditions close to the top stabilized mode ($T_k = 700K, , C_{B,k} = 5kmol/m^3$,

$C_{Char,k} = 1.7$kmol/m^3). The corresponding responses are presented in Figs. 5 and 6: (i) effluent solid apparent density and temperature as well as hot spot temperature (Fig. 5) and (ii) syngas production rate and higher heating value (Fig. 6). In terms of settling time (15–50 min), and overshoot the transient behavior of the 3-tank model resembles (within model parameter and experimental uncertainty) the ones obtained with PDE numerical solvers (Di Blasi, 2000). In particular, the predictions of the key production rate and heat content values by the 3-tank model match rather well with the ones of the 12-tank model, or equivalently, the ones obtained with respect to the previous studies with PDE numerical solvers.

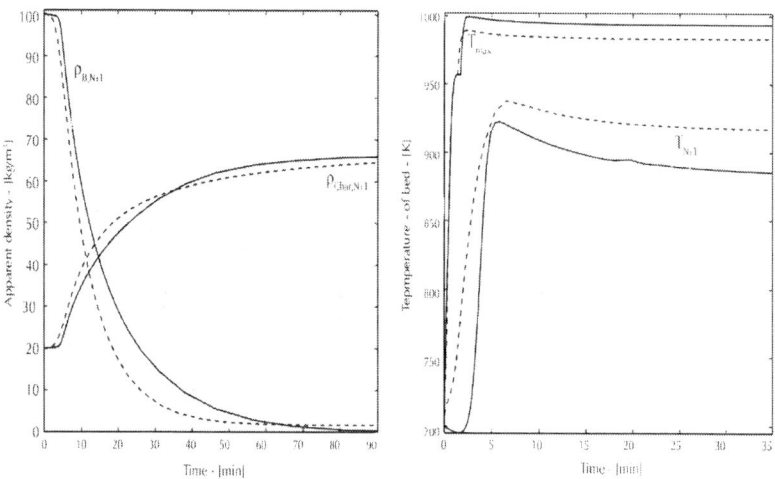

Figure 5: Effluent (biomass and char) density and temperature as well as hot spot transient responses with 3 (discontinuous plot) and 12 (continuous plot) tank models.

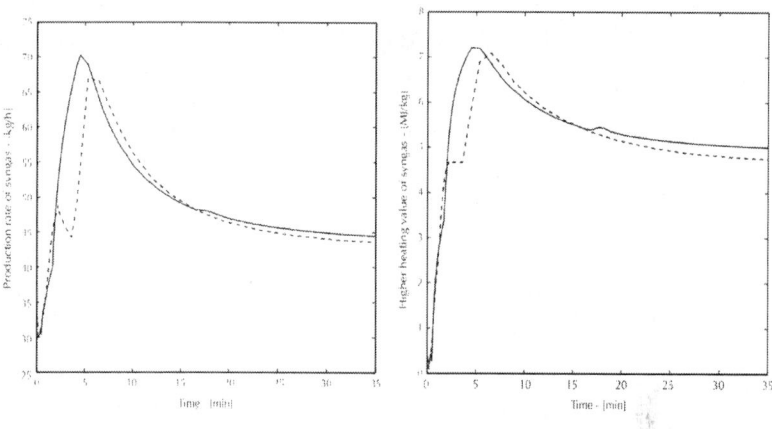

Figure 6: Transient responses of syngas production rate and higher heating value (HHV) with 3 (discontinuous plot) and 12 (continuous plot) tank models.

Summarizing, the 3-tank model yields predictions with deviations (less than 15%) similar to the ones obtained in previous simulation studies with PDE numerical solvers. Finally, a comparison between the characteristics of the 3-tank model and the ones employed in previous related modeling studies is presented in Table 7. According to Table 7 in conjunction with Table 8 the proposed modeling approach is considerably simpler in terms of the number of ODEs and of physical interpretation.

CONCLUSIONS

A reduced-order model design method suitable for (equipment and/or operation) process, monitoring and control design of moving-bed gasification reactors has been developed. The dynamic behavior of the distributed tubular gasification reactor can describe with a prescribed accuracy, by an N-CSTR model with: (i) forward and back-mixing flows that account for convective and diffusion transport, (ii) QSS gas phase assumption, and (iii) stoichiometrically-based choice of concentration invariants (or combined) components with transport reaction (or just transport) mechanism. The number N of tanks and their volumes were regarded as design degrees of freedom that are chosen according to the specific modeling purpose, in the light of the model parameter and measurements uncertainty.

The proposed approach was applied to a downdraft biomass-air fixed bed gasifier case example, achieving a significant model simplification: a 3-tank (9 ODEs) model with different volumes. The resulting model adequately describes the dynamic and static reactor behaviors, with similar accuracy to the one obtained before with PDE numerical solvers (Di Blasi, 2000 and Rogel and Aguillon, 2006).

The proposed modeling approach opens the avenue to pursue studies on: (i) multiplicity and robust stability assessments for safe-efficient design, (ii) development of temperature measurement-based concentration observers for monitoring and/or control purposes, and (iii) design of advanced or conventional output feedback control schemes.

ACKNOWLEDGMENTS

This research was sponsored by CONACYT (Grant number 103640).

APPENDIX A. N-CSTR MODEL EQUATIONS

First CSTR ($k=1$)

$$\dot{\mathbf{w}}_{s,1} = \theta_e^{Ms}\mathbf{w}_{s,e} - \theta_1^{Ms}\mathbf{w}_{s,1} + K_M \mathcal{M}\,\mathbf{Da}\,\mathbf{r}_1 \tag{38a}$$

$$\dot{\mathbf{e}}_{s,1} = \theta_e^{Ms}\mathbf{e}_{s,e} - \theta_1^{Ms}\mathbf{e}_{s,1} \tag{38b}$$

$$\dot{\eta}_1 = \theta_e^H \eta_e - \theta_1^H \eta_1 - K_M \mathbf{B}_r^T \mathbf{Da}\,\mathbf{r}_1 + \theta_2^{b,H}[\eta_2 - \eta_1] - St_{w,1}[\eta_1 - \eta_w] \tag{38c}$$

$$0 = \theta_e^{Mg} \mathbf{w}_{g,e} - \theta_1^{Mg} \mathbf{w}_{g,1} + \mathcal{M} \, \mathbf{Da} \, \mathbf{r}_1 + \theta_2^{b,Mg} [\mathbf{w}_{g,2} - \mathbf{w}_{g,1}] \tag{38d}$$

$$0 = \theta_e^{Mg} \mathbf{e}_{g,e} - \theta_1^{Mg} \mathbf{e}_{g,1} + \theta_2^{b,Mg} [\mathbf{e}_{g,2} - \mathbf{e}_{g,1}] \tag{38e}$$

Interior CSTRs ($2 \leq k \leq N-1$)

$$\dot{\mathbf{w}}_{s,k} = \theta_{k-1}^{Ms} \mathbf{w}_{s,k-1} - \theta_k^{Ms} \mathbf{w}_{s,k} + K_M \mathcal{M} \, \mathbf{Da} \, \mathbf{r}_k \tag{38f}$$

$$\dot{\mathbf{e}}_{s,k} = \theta_{k-1}^{Ms} \mathbf{e}_{s,k-1} - \theta_k^{Ms} \mathbf{e}_{s,k} \tag{38g}$$

$$\dot{\eta}_k = \theta_{k-1}^{H} \eta_{k-1} - \theta_k^{H} \eta_k - K_M \mathbf{B}_r^T \mathbf{Da} \, \mathbf{r}_k + \theta_{k+1}^{b,H} \eta_{k+1} - \theta_k^{b,H} \eta_k - St_{w,k}[\eta_k - \eta_w] \tag{38h}$$

$$0 = \theta_{k-1}^{Mg} \mathbf{w}_{g,k-1} - \theta_k^{Mg} \mathbf{w}_{g,k} + \mathcal{M} \, \mathbf{Da} \, \mathbf{r}_k + \theta_{k+1}^{b,Mg} \mathbf{w}_{g,k+1} - \theta_k^{b,Mg} \mathbf{w}_{g,k} \tag{38i}$$

$$0 = \theta_{k-1}^{Mg} \mathbf{e}_{g,k-1} - \theta_k^{Mg} \mathbf{e}_{g,k} + \theta_{k+1}^{b,Mg} \mathbf{e}_{g,k+1} - \theta_k^{b,Mg} \mathbf{e}_{g,k} \tag{38j}$$

Last CSTR ($k=N$)

$$\dot{\mathbf{w}}_{s,N} = \theta_{N-1}^{Ms} \mathbf{w}_{s,N-1} - \theta_N^{Ms} \mathbf{w}_{s,N} + K_M \mathcal{M} \, \mathbf{Da} \, \mathbf{r}_N \tag{38k}$$

$$\dot{\mathbf{e}}_{s,N} = \theta_{N-1}^{Ms} \mathbf{e}_{s,N-1} - \theta_N^{Ms} \mathbf{e}_{s,N}$$

(38l)

$$\dot{\eta}_N = \theta_{N-1}^{H} \eta_{N-1} - \theta_N^{H} \eta_N - K_M \mathbf{B}_r^T \mathbf{Da}\ \mathbf{r}_N - \theta_N^{b,H} \eta_N - St_{w,N}[\eta_N - \eta_w]$$

(38m)

$$0 = \theta_{N-1}^{Mg} \mathbf{w}_{g,N-1} - \theta_N^{Mg} \mathbf{w}_{g,N} + \mathcal{M}\ \mathbf{Da}\ \mathbf{r}_N - \theta_N^{b,Mg} \mathbf{w}_{g,N}$$

(38n)

$$0 = \theta_{N-1}^{Mg} \mathbf{e}_{g,N-1} - \theta_N^{Mg} \mathbf{e}_{g,N} - \theta_N^{b,Mg} \mathbf{e}_{g,N}$$

(38o)

where

$$\boldsymbol{\theta}_e = \frac{\mathbf{q}_e}{V_1}, \quad \boldsymbol{\theta}_k = \frac{\mathbf{q}_k}{V_k}, \quad \boldsymbol{\theta}_k^b = \frac{\mathbf{q}_k^b}{V_k}$$

(38p)

$$\boldsymbol{\theta}_e = \begin{bmatrix} \theta_e^{Ms} \\ \theta_e^{H} \\ \theta_e^{Mg} \end{bmatrix}, \quad \boldsymbol{\theta}_k = \begin{bmatrix} \theta_k^{Ms} \\ \theta_k^{H} \\ \theta_k^{Mg} \end{bmatrix}, \quad \boldsymbol{\theta}_k^b = \begin{bmatrix} \theta_k^{b,Ms} \\ \theta_k^{b,H} \\ \theta_k^{b,Mg} \end{bmatrix}$$

(38q)

$$V_R = LA_R = \sum_{k=1}^{N} V_k, \quad V_k = \Delta z_k A_R$$

(38r)

Vk (or VR) is the volume of the k -th tank (or tubular reactor), $_{zk}$ is the distance between k -th node and k−1-th node of the spatial discretization mesh, AR is the reactor section area, $_e$ is the feed dilution rate vector associated with the feed flow vector $_{qe}$ with solid mass q_e^{Mg}, heat q_e^{H} and gas mass q_e^{Mg} components, and $_k$ (or θ_k^b is the forward (or back-mixing) dilution rate vector associated to the volumetric flow vector $_{qk}$ (or θ_k^b with solid mass, heat and gas mass forward (or back-mixing) flows q_k^{Mg}, q_k^{Mg} and q_k^{Mg} (or q_k^{bMs}, q_k^{bH} and q_k^{bMs}, respectively.

REFERENCES

1.   Amundson, N.R., Arri, L.E., 1978. Char gasification in a countercurrent reactor. AIChE Journal, 87–101.

2.   Aris, R., 1961. The Optimal Design of Chemical Reactors: A Study in Dynamic Programming. Academic Press, N.Y.

3.   Aris, R., 1965. Introduction to the Analyses of Chemical Reactors. Prentice-Hall, New Jersey.

4.   Baldea, M., Daoutidis, P., 2007. Dynamics and control of autothermal reactors for the production of hydrogen. Chemical Engineering Science 62, 3218–3230.

5.   Barrio, M., Fossum, M., Hustad, J., 2001. A small-scale stratified downdraft gasifier coupled to a gas engine for combined heat and power production. Progress in Thermochemical Biomass Conversion, Blackwell Science 1, 426–440.

6.   Basu, P., 2006. Combustion and Gasification in Fluidized Beds. Taylor and Francis. Beér, J.M., 2007. High efficiency electric power generation: the environmental role. Progress in Energy and Combustion Science, 107–134.

7.   de Boor, C., 2001. A Practical Guide to Splines. Springer. Caram, H.S., Fuentes, C., 1982. Simplified model for a countercurrent char gasifier. Industrial and Engineering

Chemistry Fundamentals 21, 464–472.

8.   Castellanos-Sahagun, E., Alvarez, J., Alvarez-Ramirez, J., 2006. Two-point composition-temperature control of binary distillation columns. Industrial and Engineering Chemical Research 45, 9010–9023.

9.   Chen, W.-H., Chen, J.-C., Tsai, C.-D., Jiang, T.L., 2007. Transient gasification and syngas formation from coal particles in a fixed-bed reactor. International Journal of Energy Research 31, 895–911.

10.   Coste, J., Rudd, D., Amundson, N.R., 1961. Taylor diffusion in tubular reactors. The Canadian Journal of Chemical Engineering 39, 149–151.

11.   Deans, H.A., Lapidus, L., 1960. A computational model predicting and correlating the behavior of fixed bed reactors: II extension to chemically reactive systems. AIChE Journal 6, 663–668.

12.   Di Blasi, C., 2000. Dynamic behaviour of stratified downdraft gasifiers. Chemical Engineering Science 55, 2931–2944.

13.   Di Blasi, C., 2004. Modeling wood gasification in a countercurrent fixed-bed reactor. AIChE Journal 50, 2306–2319.

14.   Diaz-Salgado, J., Alvarez, J., Schaum, A., Moreno, J., 2012. Feedforward outputfeedback control for continuous exothermic reactors with isotonic kinetics. Journal of Process Control 22, 303–320.

15.   Feinberg, M., 1977. Mathematical aspects of mass action kinetics. In: Lapidus, Leon, Amundson, Neal R. (Eds.), Chemical Reactor Theory, A Review. Prentice Hall. (Chapter 1).

16.   Fernandez, C., Alvarez, J., Baratti, R., Frau, A., 2012. Estimation structure design for staged systems. Journal of Process Control 22, 2036–2056.

17.   Gobel, B., Henriksen, U., Jensen, T.K., Qvale, B., Houbak, N., 2007. The development of a computer model for a fixed bed gasifier and its use for optimization and control. Bioresource

Technology 98, 2043–2052.

18. Grieco, E.M., Baldi, G., 2011. Predictive model for countercurrent coal gasifiers. Chemical Engineering Science 66, 5749–5761.

19. Higman, C., van der Burgt, M., 2008. Gasification, (2nd ed.) Gulf Professional Publishing.

20. Krstic, M., Kanellakopoulos, I., Kokotovic, P., 1995. Nonlinear and Adaptive Control Design. Wiley, New York.

21. Lopez, T., Alvarez, J., 2004. On the effect of the estimation structure in the functioning of a nonlinear copolymer reactor estimator. Journal of Process Control 14, 99–109.

22. Manurung, R., Beenackers, A., 1993. Modeling and simulation of an open core downdraft moving bed rice husk gasifier. Proceedings of the International Conference on Advances in Thermochemical Biomass Conversion 1, 288–309.

23. Martinez-Vera, C., Ruiz-Martinez, R., Vizcarra-Mendoza, M., Alvarez-Calderon, J., 2010. Apparent diffusion model assessment in extraction processes by means of a Luenberger observer. Journal of Food Engineering 101, 16–22.

24. Najera, I., 2012. Modeling and control of a class of exothermic tubular reactors (in Spanish). Master Degree Thesis, UAM Iztapalapa, Mexico, D.F. Reed, T., Walt, R., Ellis, S., Das, A., Deutch, S., 1999. Superficial velocity—the key to downdraft gasification. Fourth Biomass Conference of the Americas, Oakland, USA 25, 343–356.

25. Reed, T.B., Das, A., 1988. Handbook of Biomass Downdraft Gasifier Engine Systems. Solar Technical Information Program. Reed, T.B., Markson, M., 1985. Biomass gasification reaction velocities. Fundamentals of Thermochemical Biomass Conversion. Elsevier, p. 951.

26. Rogel, A., Aguillon, J., 2006. The 2d Eulerian approach of entrained flow and temperature in a biomass stratified downdraft gasifier. American Journal of Applied Sciences 3, 2068–2075.

27. Rogel, A.R., 2007. Estudio numerico experimental de un gasificador estratificado que opera con biomasa, utilizando CFD. Tesis de Doctorado, Programa de Maestría y Doctorado en Ingeniería UNAM, Mexico, D.F.

28. Schaum, A., Alvarez, J., Lopez-Arenas, T., 2012. Output-feedback saturated control of a class of continuous biological reactors with inhibited kinetics. Chemical Engineering Science 68, 520–529.

29. Sepulchre, R., Jankovic, M., Kokotovic, P., 1997. Constructive Nonlinear Control. Springer.

30. Sheth, P.N., Babu, B., 2009. Experimental studies on producer gas generation from wood waste in a downdraft biomass gasifier. Bioresource Technology 100, 3127–3133.

31. Shwe, S., 2004. A theoretical and experimental study on a stratified downdraft biomass gasifier. PhD Thesis, University of Melbourne.

32. Varma, A., Amundson, N.R., 1973. Some observations on uniqueness and multiplicity of steady states in non-adiabatic chemically reacting systems. The Canadian Journal of Chemical Engineering 51, 206.

33. Zainal, Z., Rifau, A., Quadir, G., Seetharamu, K., 2002. Experimental investigation of a downdraft biomass gasifier. Biomass and Bioenergy 23, 283–289

6

A New Method to Calculate Kinetic Parameters Independent of the Kinetic Model: Insights on CO_2 and Steam Gasification

Arturo Gomez and Nader Mahinpey

Dept. of Chemical and Petroleum Engineering, Schulich School of Engineering, University of Calgary, 2500 University Drive NW, Calgary, AB T2N 1N4, Canada

ABSTRACT

A new method to obtain the rate constant and activation energy independent of a kinetic model is proposed and evaluated for thermochemical conversion, specifically in the steam and CO_2 gasification of coal and biomass. Recent works on gas–solid reactions are based on single-step chemical reaction models that have been increasing in complexity through the use of more regression parameters to fit experimental data. These models fit better; however, sometimes their kinetic parameters are inconsistent, resulting in an incorrect interpretation of the reaction mechanism.

The proposed method, which does not require any assumed kinetic model, is useful in calculating the parameters of the Arrhenius equation using cumulative variables obtained from the experimental data, i.e. conversion and residence time. For this reason, the uncertainty is reduced compared to conventional methods. The new method could be used as a consistency test between different kinetic models by comparing their kinetic parameters with those obtained with the proposed free-model method.

The procedure has been applied to our previous experimental work and other authors' information on CO_2 and steam gasification, verifying that the random pore model is not the best kinetic model to represent gasification and partial oxidation of coal and biomass. The new procedure can be used as a tool for chemical reaction engineering analysis in a broad range of thermochemical reactions under isothermal consideration.

Graphical Abstract

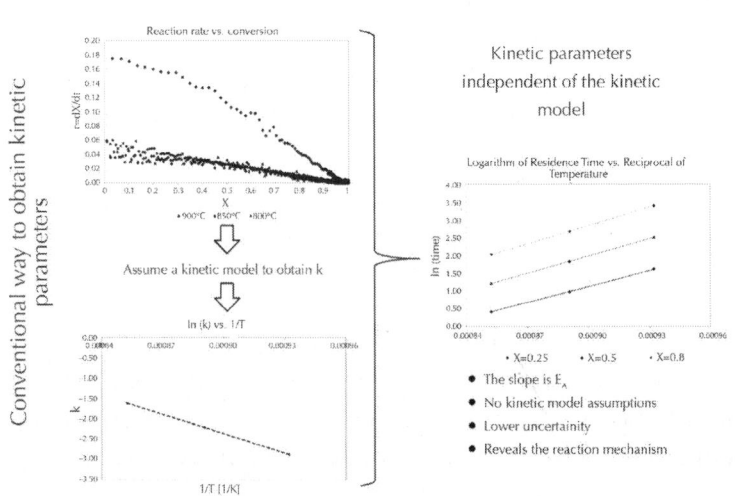

INTRODUCTION

Gasification is one of the most promising thermochemical conversion technologies to use alternative fuels as feedstock, especially low-rank coals and biomass. Reviews have presented the most significant gasification variables (Di Blasi, 2009, Irfan et al., 2011 and Hobbs et al., 1993); however, the kinetics and analysis of its reaction mechanism are complex, since the reaction occurs at high temperatures and the solid characterization usually is performed at very low temperatures.

Kinetic information of partial oxidation and combustion has been reported, since the most common industrial gasifiers inject air to partially combust the fuel, providing the energy that the overall endothermic process requires. Moreover, experiments and modeling for partial oxidation (Loewenberg and Levendis, 1991 and Su and Perlmutter, 1985) have been extended for gasification modeling. Studies on gasification have been performed in carbon dioxide (CO_2) (Duman et al., 2014, Jeong et al., 2014, Li et al., 2013,Mandapati et al., 2012, Silbermann et al., 2013 and Wang et al., 2013), steam atmospheres (Fermoso et al., 2011, Kim et al., 2013 and Lin and Strand, 2014), or mixtures of both gasifying agents (Ahmed and Gupta, 2011, Guizani et al., 2013, Ren et al., 2013, Umemoto et al., 2013 and Zhang et al., 2014), since the Boudouard reaction, steam reforming and water–gas shifting are the main reactions. Analysis of the reported data in this field is complex, since there is not a criterion consensus (Di Blasi, 2009); and, the modeling of a maximum reaction rate has been the focus of the research on gasification kinetics in recent years (Bhatia and Perlmutter, 1980, Bhatia and Vartak, 1996, Duman et al., 2014, Jeong et al., 2014,Kopyscinski et al., 2013, Li et al., 2013, Lin and Strand, 2014, Mandapati et al., 2012 and Singer and Ghoniem, 2011).

Bhatia and Perlmutter (1980) proposed the random pore model (RPM) with a further modification (Bhatia and Vartak, 1996) for gas–solid reactions. This model has been widely accepted due to its nonlinear dependence on char surface, which can predict a maximum reaction rate as observed experimentally. Different modifications to the original model and their applications to fit experimental data have been reported; for example, some of the most recent works present extended and adaptive RPM (Kopyscinski et al., 2013 and Singer and Ghoniem, 2011). Modeling improvement is commonly attached to an increase in the number of the fitting parameters, which does not necessarily mean a direct relationship with the reaction mechanism.

Recently, Gomez et al. (2014) demonstrated that the suggested maximum rate is a consequence of a change in the reaction medium, which is generated by an imposition of the experimental procedure, and proposed an alternative experimental method to avoid this effect. In independent studies (Ahmed and Gupta, 2011, Li et al., 2013, Nipattummakul et al., 2010, Popa et al., 2013, Prabowo et al., 2014 and Woodruff and Weimer, 2013), the time to observe a maximum rate was constant and independent of the char sample or gasifying agent, as proven by Gomez et al., despite many authors modeled this maximum (Ahmed and Gupta, 2011, Li et al., 2013 and Popa et al., 2013). For this reason, simpler expressions can be used to model gasification or other thermochemical reactions where the reaction is chemically controlled and thus one single overall step can be assumed. Therefore, it is important to validate the assumed kinetic model and its respective kinetic parameters (i.e. rate constant and activation energy).

A new procedure is presented to obtain the rate constant and activation energy, based on a deduction from the Arrhenius equation and a general rate law, without transformation of variables or assumption of a particular kinetic model. The aim of this work is the determination of kinetic parameters without restricting the analysis to a particular kinetic model. From reported data for CO_2 (Kopyscinski et al., 2013, Li et al., 2013, Mandapati et al., 2012 and Silbermann et al., 2013) and steam gasification (Fermoso et al., 2011), the activation energy was calculated with the new approach and compared with the reported values, confirming previous findings (Gomez et al., 2014) related to the convenience of using simpler models rather than the RPM for gasification. This new procedure can be used to determine the parameters of the Arrhenius equation for a set of isothermal experiments and can also be used as a tool for scaling industrial processes or testing the consistency of a particular kinetic model.

EXPERIMENTAL METHODS

CO_2 Gasification

Original experimental information from Silbermann et al. related to CO_2 coal gasification was used to determine the activation energy and compare the obtained values with those reported for five different kinetic models (Silbermann et al., 2013). The same procedure was applied to three other works using a nonlinear model (Li et al., 2013) and to the RPM (Kopyscinski et al., 2013 and Mandapati et al., 2012). They reported their results as the best fit among the compared kinetic models. It is important to mention that the main experimental difference between Silbermann et al. (2013) and the other references is that its experimental procedure did not induce a maximum rate as a consequence of a gas change, as proven byGomez et al. (2014).

Steam Gasification

Results for CO_2 and steam gasification follow the same trend, with a higher reactivity of the steam at lower temperatures. Kinetic modeling for steam gasification, using a single-step chemical reaction model, is similar to that of CO_2 gasification (Ahmed and Gupta, 2011). When CO_2 and steam are mixed in different proportions, Langmuir–Hinshelwood (LH) models describe the competition for active sites considering the gas diffusion (Umemoto et al., 2013), but the chemical reaction contribution are assumed with a single-step kinetic model. Information presented by Fermoso et al. (2011) was analyzed in the application of the proposed method to determine the activation energy and compare it with the reported values obtained using the RPM.

KINETICS ANALYSIS

Data Analysis

Conversion and its associated reaction time were obtained from five independent studies; i.e. Fermoso et al. (2011), Kopyscinski et al. (2013), Li et al. (2013), Mandapati et al. (2012) and Silbermann et al. (2013). Conversion is calculated from the weight at a particular time, which is the original information obtained by thermogravimetic analysis (TGA) or back calculating the information of the gas composition analysis. By definition, conversion is:

$$X = \frac{m_0 - m_t}{m_0 - m_a}$$

(1)

where m_o is the initial mass of the sample, m_t is the mass at a particular time, and ma is the mass of the ash.

The conversion rate was not determined in this work, since the proposed method does not require it. This approach is especially useful, as many works have reported a maximum, which can affect the accuracy of the kinetic parameters. Considering the solid molar balance in a batch reactor, the conversion rate, r, is expressed as:

$$r = \frac{dX}{dt}$$

(2)

Regression parameters in all references were calculated with least square minimization and compared using the coefficient of determination (R^2). A complete description of the most common kinetic models can be found elsewhere (Silbermann et al., 2013). In this work, linearized or linear regression of logarithm expressions were used to obtain these parameters. The activation energy was estimated through the correlation of data at different reaction temperatures in an alternative form of the Arrhenius equation.

Calculation of the Activation Energy from Experimental Data

A general rate law for gas–solid reactions can be represented as the product of two independent functions of the independent variables' temperature and conversion under isobaric considerations. Without considering the effect of the catalyst and mass transfer limitations, the conversion rate is given by:

$$r = \frac{dX}{dt} = k(T)f(X)$$

(3)

where k is the rate constant and $f(X)$ is a function of the solid surface and usually associated with the solid conversion.

The effect of temperature on the reaction rate is well described by the Arrhenius equation, which is given by:

$$k = k_0 e^{-\frac{E_A}{RT}}$$

(4)

where EA is the activation energy (kJ/mol), ko is the frequency factor (min^{-1}), R is the ideal gas law constant (kJ/mol K), and T is the absolute temperature (K).

Solving the first-order differential of Eq. (3), the product of time for the rate constant is a constant for a particular conversion:

$$\int_0^X \frac{dX}{f(X)} = G(X) = k_X(T)t_X(T)$$

(5)

where k_x and t_x are the rate constant (min^{-1}) and residence time (min) for a fixed conversion (X), and both are functions of the temperature. From the general rate law and the Arrhenius equation, i.e. Eqs. (5) and (3), respectively, it is possible to state the following correlation of variables:

$$k_X(T) \sim \frac{1}{t_X(T)} \sim e^{-\frac{E_A}{RT}}$$

(6)

or in a logarithmic form:

$$\ln[t_X(T)] \sim -\ln[k_X(T)] \sim \frac{E_A}{RT}$$

(7)

The formal solution of Eq. (5) leads to:

$$\ln[t_X(T)] = \{\ln[G(X)] - \ln[k_o]\} + \frac{E_A}{RT} = \alpha + \frac{E_A}{RT}$$

(8)

where $\alpha = \ln[G(X)] - \ln[k_o]$ and is a constant for a particular conversion.

From Eq. (8), it is evident that the ratio of the activation energy and the ideal gas law constant (E_A/R) is the slope of the logarithm of time versus the reciprocal of temperature. By definition, k is independent of the conversion in Eq. (3); thus, the plot $\ln(t)$ versus $1/T$ for different conversions should exhibit a linear trend and the same slope if the reaction mechanism follows the same path in the whole temperature range.

One of the limitations of kinetic analysis is the selection of the conversion to obtain the rate constant and other parameters involved in a particular model. Many authors use conversions lower than 100%, since uncertainty increases as the sample weight approaches zero. For this reason, the proposed method has just one way to obtain the activation energy, with a particular conversion representing the whole conversion range for practical purposes.

An alternative procedure to determine the activation energy without a kinetic model is using the initial reaction rate, which should be exactly the same rate constant for almost all kinetic models. To understand this, consider the simple power law kinetic model or integrated core model (ICM) as a particular case of Eq.(3):

$$r = \frac{dX}{dt} = k(1 - X)^n \to k = \frac{dX}{dt}\bigg]_{X \to 0}$$

(9)

The main limitation of Eq. (9) is related to the accuracy of the initial reaction rate determination, when an inert gas is switched to a gasifying agent. Another limitation is that the calculation of the initial rate requires continuous reading of the weight variation (or small intervals of time), which can be difficult to achieve with techniques other than TGA (i.e. gas chromatography analysis). The most important application of Eq. (9) is for the consistency evaluation of a particular kinetic model, since the initial reaction rate should be close to the rate constant.

The uncertainty of the activation energy calculated as proposed in this study is smaller than the one reported by the independent studies considered as study cases, as there is uncertainty propagation of k with a kinetic model since this value is obtained from a regression for each isothermal experiment. Using the new approach, just one regression is necessary, instead of $m+1$ with the Arrhenius equation, i.e. one regression to obtain k (for m different temperatures) plus the Arrhenius equation. The maximum uncertainty for the activation energy estimated with the proposed method, considering three temperatures and two repetitions, is ±13.5 kJ/mol with a coefficient of determination higher than 0.99. It is important to mention that residence time is a cumulative function and increases monotonically with conversion, thereby improving accuracy by increasing conversion.

Frequency Factor Approximation

As mentioned in the analysis of Eq. (9), the rate constant should be equal to the initial reaction rate when the conversion is close to zero. In fact, the rate constant is the most difficult parameter to be determined, since the experimental procedure affects the initial rate and the partial pressure of the gasifying agent is often not constant during the first instant of the reaction (Gomez et al., 2014).

A similar expression to Eq. (8) was presented by De Micco et al. (2012), but there was no reference about the frequency factor estimation independent of the kinetic model. From Eq. (8), the intercept at constant conversion ( ) is a term including conversion and the rate constant at infinite temperature. It is expressed in the following equation:

$$\ln[k_0] = \ln[G(X)] - \alpha$$

(10)

where $G(X) = \int_0^x (dx / f(x))$ is a function for a particular conversion, k_o (min^{-1}) is the rate constant at infinite temperature or frequency factor, and (min^{-1}) is the intercept of Eq. (8). Function $G(X)$ is unknown; however, it is possible to infer the magnitude order between and $G(X)$. For example, for an n-order reaction model (ICM), this function is given by:

$$G(X)_{n\,order} = \begin{cases} \ln\left(\dfrac{1}{1-X}\right) & \text{if } n = 1 \\[2mm] \dfrac{(1-X)^{1-n} - 1}{n - 1} & \text{if } n \neq 1 \end{cases}$$

(11)

Table 1 shows the values of $G(X)$ and its logarithm at different conversion based on Eq. (11). The respective values at 80% and 50% conversion are highlighted; illustrating that logarithm of $G(X)$ is negative below 50% conversion and positive above 80% conversion regardless of the reaction order. The frequency factor for steam and CO_2 gasification is in the range of 1×10^4 min^{-1} to 1×10^{10} min^{-1} between 700 °C and 900 °C (Di Blasi, 2009, Fermoso et al., 2011, Irfan et al., 2011 and Silbermann et al., 2013), where the reaction is chemically controlled. The logarithm of k_o (min^{-1}) in the mentioned range is between 9.2 and 23. For a conversion range between 0.5 and 0.8, the absolute value of the function $G(X)$ is much smaller than ; therefore, the following expression can be considered for CO_2 and steam gasification:

$$\ln[k_0] = \ln[G(X)] - \alpha \approx -\alpha \quad 0.5 \leq X \leq 0.8, \quad T \gg 700\,°C$$

(12)

Table 1: Analytic values of $G(X)$ according to Eq. (11) and their respective logarithms for three reaction orders: 0.5, 1 and 2

X	G(X) at the indicated reaction order			ln[G(X)]		
	0.5	1	2	0.5	1	2
0.00	0.00	0.00	0.00	−6.91	−6.91	−6.91
0.01	0.01	0.01	0.01	−4.60	−4.60	−4.60
0.05	0.05	0.05	0.05	−2.98	−2.97	−2.94
0.10	0.10	0.11	0.11	−2.28	−2.25	−2.20
0.20	0.21	0.22	0.25	−1.56	−1.50	−1.39
0.30	0.33	0.36	0.43	−1.12	−1.03	−0.85
0.40	0.45	0.51	0.67	−0.80	−0.67	−0.41
0.50	0.59	0.69	1.00	−0.53	−0.37	0.00
0.60	0.74	0.92	1.50	−0.31	−0.09	0.41
0.70	0.90	1.20	2.33	−0.10	0.19	0.85
0.80	1.11	1.61	4.00	0.10	0.48	1.39
0.90	1.37	2.30	9.00	0.31	0.83	2.20
0.95	1.55	3.00	19.00	0.44	1.10	2.94
0.99	1.80	4.61	99.00	0.59	1.53	4.60
1.00	1.94	6.91	999.00	0.66	1.93	6.91

Bold values signify the upper and lower limits of G(X) to determine the frequency factor using Eq. (8).

A simple way to determine k_o is the use of the average of for 2 different conversions, if the absolute difference between both intercepts is lower than 1 min^{-1}. Although this is not the exact value, it does give a very good idea about the magnitude order of the frequency factor; and considering 50% and 80% conversion provides the upper and lower limits of the frequency factor. For other reactions, the same analysis applies with special attention to the selected temperature range.

RESULTS AND DISCUSSION

CO_2 Gasification Kinetics from Alberta Coals

The original data reported by Silbermann et al. (2013) were time and weight loss; and, the activation energy and other kinetic parameters (depending on the kinetic model) were obtained at an 80% conversion. Conversions and the associated times for nine different coal samples (seven deep coals and two surface mined coals) are presented in Table 2.

Table 2: Time (min) to reach three different conversions ($X_1 = 0.8$; $X_2 = 0.5$; $X_3 = 0.25$) for nine different coals. CO_2 gasification at 800 C, 850 C and 900 C. Reported by Silbermann et al. (2013)

T [°C]	Residence time (min) at different temperature (°C)								
	$X = 0.8$			$X = 0.5$			$X = 0.25$		
	800	850	900	800	850	900	800	850	900
Genesee 1	29.8	13.2	5	11.4	5.2	2.1	4.2	1.9	0.85
Genesee 2	49.6	19.8	7.4	17.2	6.8	2.8	6	2.4	1.15
Coal 5	195.8	66.2	20.8	96.2	34.6	11.6	35.6	14	5.1
Coal 2	117.2	45.6	21.8	28.6	10.8	5.5	7.2	3	1.8
Coal 3	284.2	86.2	24.8	126.6	34.9	12.4	45.6	15.3	5.2
Coal 4	362.6	102.6	49	181.6	52.2	19.4	60.4	18.4	7.1
Coal 6	530.4	162.6	57.6	92.6	26.6	11.4	24.6	7.5	3.5
Coal 1	278.6	148.2	62.6	74	39.8	16.4	19.4	10	4.3
Coal 7	799.2	220	80.1	352	88.6	32.6	117.6	32.4	13.6

From Eq. (8), the logarithm of time versus the reciprocal of temperature was plotted in Fig. 1 for two of the nine coals presented by Silbermann et al. (2013). It is worth noting that all nine coals exhibited the same trends; however, for practical reasons, just the coals with the fastest (Genesee) and the slowest (coal 7) reactivity are shown in Fig. 1. The slopes of the logarithm of time versus the reciprocal of temperature plots followed a linear trend and varied slightly with conversion.

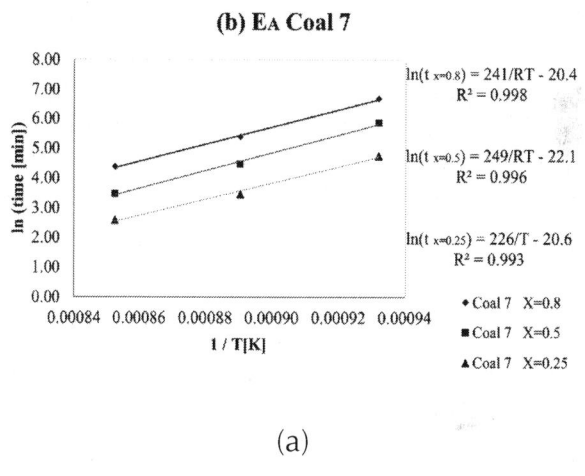

(a)

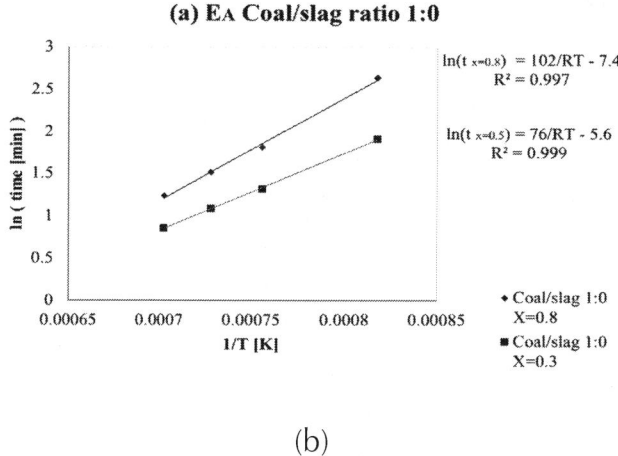

(b)

Figure 1: Logarithm of time versus reciprocal of temperature. CO_2 gasification between 800 C and 900 °C: (a) Genesee coal and (b) Deep coal 7. Data from Silbermann et al. (2013).

Complete information of the activation energy for all nine coals is shown in Table 3. The activation energies obtained by regression of the Arrhenius equation for the first four models (ICM, VM, SCM and NDM) were close to the activation energy calculated from Eq. (8) (absolute deviation was smaller than 20 kJ/mol). Using the proposed method as the reference value, the deviation of the RPM was the highest among all five models. For example for coal 5, the activation energy estimated with the RPM (140 kJ/mol) is lower than the other models and the free-model approach (higher than 200 kJ/mol); it is evident that the RPM did not represent the reaction mechanism of coal 5.

Table 3: Activation energy [kJ/mol] based on Eq. (8) and the Arrhenius equation using five different kinetic models as reported bySilbermann et al. (2013). Temperature range from 1073 K to 1173 K

	EA (kJ/mol) from Eq. (8)			EA (kJ/mol) for each kinetic model at $X = 0.8$				
	$X = 0.8$	$X = 0.5$	$X = 0.25$	VM	SCM	IM	NDM	RPM
Genesee 1	186	177	167	180	183	172	175	184
Genesee 2	199	192	173	191	187	193	195	187
Coal 1	156	157	157	124	119	139	137	117
Coal 2	176	173	146	139	128	164	169	126
Coal 3	255	243	227	209	209	208	216	211
Coal 4	210	220	205	162	153	209	205	151
Coal 5	234	221	203	203	205	193	200	140
Coal 6	232	234	224	186	186	209	211	187
Coal 7	241	249	226	205	209	230	233	212

The ICM yielded the most accurate results among all models, and a similar result was reported (Irfan et al., 2011) through the analysis of the coefficient of determination. The reasons were the lack of a maximum reaction rate and a logarithmic correlation that better fit the experimental data. It was shown that the RPM underestimated the activation energy; therefore, there is no reason to consider such a complex model in direct gasification, i.e. no gases switching.

A comparison of the reference activation energies from 50% and 80% conversions, as obtained from Eq.(8), indicated that they were similar, given that the average relative uncertainty was ±7.3%. The maximum difference between coals is expressed as:

$$\text{Maximum} \left[\frac{\left| E_{A,X=0.8} - E_{A,X=0.5} \right|}{E_{A,X=0.8}} \right] = 5.2\%, \quad \text{for coal 5}$$

The same comparison between 25% and 80% conversions yielded a higher difference 17% for coal 2, which had the highest ash content of all coal samples. Moreover, the activation energy was smaller at the lower conversion for all cases: $E_{A,X=0.8} \geq E_{A,X=0.25}$ (columns 2–4 and 8–10 of Table 2, respectively). This may be attributed to part of the alkali being inactivated or lost during the gasification, or intraparticle diffusion when char to ash ratio is decreasing. This analysis is important since one of the conditions for intrinsic kinetic modeling is the assumption that the catalyst remains active.

Another possible reason to be considered when reactivity decreases during long time exposition at high temperature is thermal annealing (Senneca et al., 1997); however, the coals exposed to a longer gasification time (coals 7, 6 and 4) show similar activation energy between 50% and 80% conversion (within the uncertainty). If the reaction follows the same mechanism in the whole conversion range, the activation energy calculated at different conversions should be almost the same. This analysis can be useful to understanding the mechanism path of any gas–solid reaction.

The Arrhenius frequency factor calculated using Eq. (12) is presented in the 5th column of Table 4 and it was taken as the reference for comparison with the frequency factors obtained from the original data for the five different models. The ICM had the closest value to the reference; and, the RPM underestimated the frequency factor and, in some cases such as the coal 5, had a very different value when comparing their relative magnitude orders. This indicates that the calculation of ko using Eq. (12) is in good agreement with the experimental results and gives a clear indication about the magnitude order of the frequency factor.

Table 4: Intercept of Eq. (8) and frequency factor (min^{-1}) based on Eq. (12) (fourth column) and the Arrhenius equation using five different kinetic models as reported by Silbermann et al. (2013). Temperature range from 1073 K to 1173 K

	α (min^{-1}) Eq. (8)			ko	ko (min^{-1}) at 80% for each kinetic model				
	X = 0.8	X = 0.5	X = 0.25	Eq. (12)	VM	SCM	ICM	NDM	RPM
Genesee 1	29.8	13.2	5	3.6E+07	3.6E+07	3.7E+07	1.6E+07	1.9E+07	3.5E+07
Genesee 2	49.6	19.8	7.4	1.1E+08	7.7E+07	4.3E+07	1.1E+08	1.2E+08	3.5E+07
Coal 5	195.8	66.2	20.8	8.7E+08	7.0E+07	7.5E+07	1.8E+07	3.7E+07	7.1E+04
Coal 2	117.2	45.6	21.8	5.6E+06	1.7E+05	4.0E+04	3.4E+06	4.6E+06	2.7E+04
Coal 3	284.2	86.2	24.8	7.1E+09	1.1E+08	1.0E+08	8.5E+07	1.8E+08	1.0E+08
Coal 4	362.6	102.6	49	2.7E+08	9.6E+05	3.1E+05	1.9E+08	9.6E+07	1.9E+05
Coal 6	530.4	162.6	57.6	4.7E+08	6.8E+06	5.9E+06	9.1E+07	8.9E+07	5.7E+06
Coal 1	278.6	148.2	62.6	2.7E+05	1.3E+04	6.6E+03	7.7E+04	5.4E+04	4.3E+03
Coal 7	799.2	220	80.1	1.7E+09	3.2E+07	4.0E+07	4.3E+08	5.2E+08	4.8E+07

CO$_2$ Gasification Kinetics from Coals with Slag Granules

The conversion and time at different temperatures for CO$_2$ gasification of four different coal/slag ratios (1:0, 1:1, 1:2, 1:3), as reported by Li et al. (2013), were correlated using Eq. (8). Fig. 2 shows the plots of the logarithm of time versus the reciprocal of temperature for two different sets of conversion data points. The experimental procedure performed in the original study switched gases after reaching the reaction temperature. The temperature range from 1223 K to 1423 K was the same for all four different coal/slag samples.

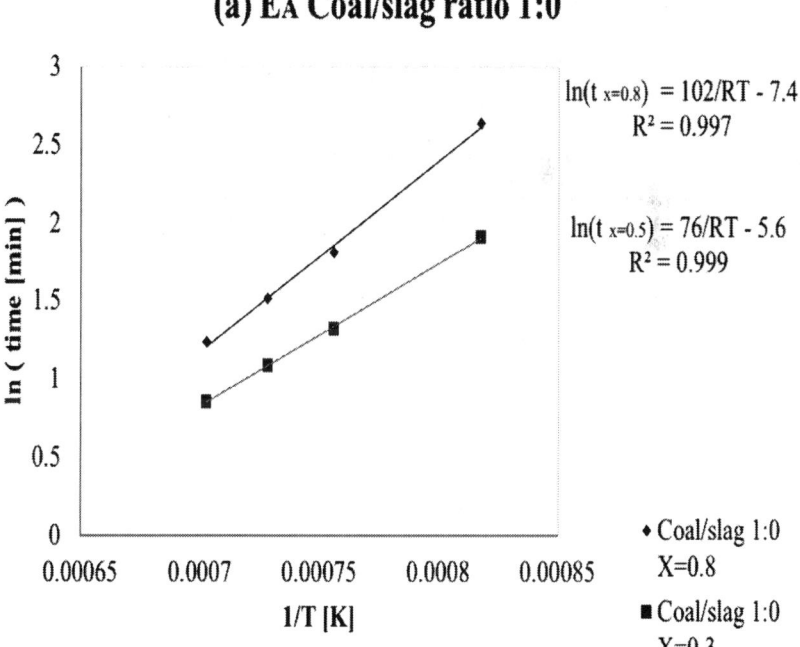

(a) E$_A$ Coal/slag ratio 1:0

$\ln(t_{x=0.8}) = 102/RT - 7.4$
$R^2 = 0.997$

$\ln(t_{x=0.5}) = 76/RT - 5.6$
$R^2 = 0.999$

◆ Coal/slag 1:0 X=0.8
■ Coal/slag 1:0 X=0.3

(a)

(b) Eᴀ Coal/slag ratio 1:1

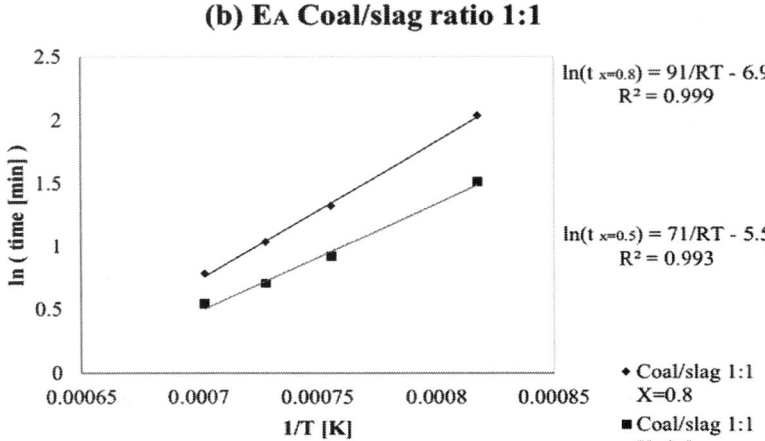

$\ln(t_{x=0.8}) = 91/RT - 6.9$
$R^2 = 0.999$

$\ln(t_{x=0.5}) = 71/RT - 5.5$
$R^2 = 0.993$

♦ Coal/slag 1:1 X=0.8
■ Coal/slag 1:1 X=0.5

(b)

(c) Eᴀ Coal/slag ratio 1:2

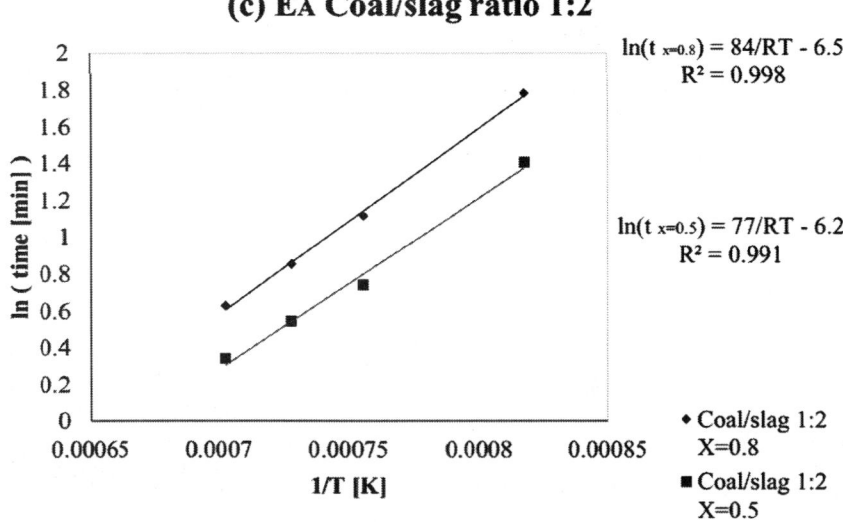

$\ln(t_{x=0.8}) = 84/RT - 6.5$
$R^2 = 0.998$

$\ln(t_{x=0.5}) = 77/RT - 6.2$
$R^2 = 0.991$

♦ Coal/slag 1:2 X=0.8
■ Coal/slag 1:2 X=0.5

(c)

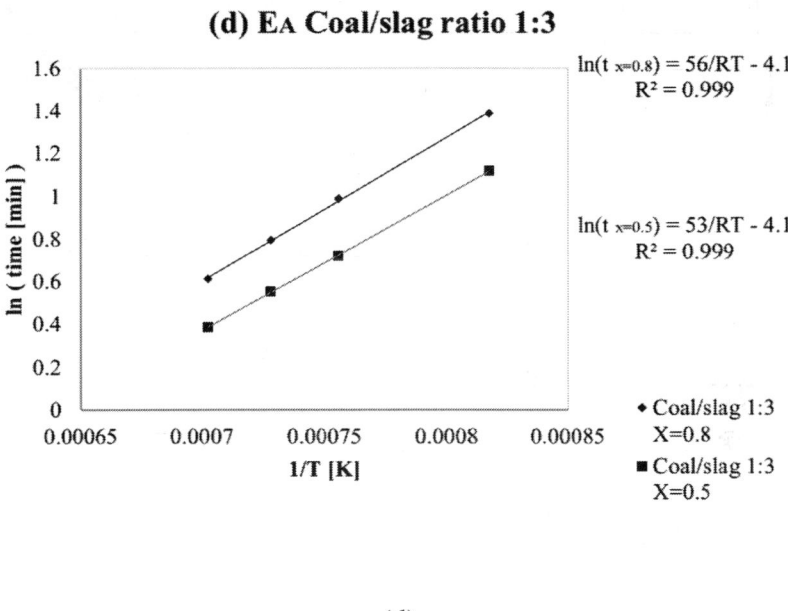

(d)

Figure 2: Logarithm of time versus reciprocal of temperature. CO_2 gasification between 950 °C and 1150 °C: (a) coal/slag ratio of 1:0, (b) coal/slag ratio of 1:1, (c) coal/slag ratio of 1:2, and (d) coal/slag ratio of 1:3. Data from Li et al. (2013).

Table 5 shows the reference activation energies and frequency factors obtained using Eqs. (8) and (12), respectively, for 80% and 50% conversions. The same parameters were reported using a nonlinear kinetic model, i.e. the Avrami–Erofeev ($m = 2$) model (Li et al., 2013), as presented in columns 4 and 7 of Table 5, respectively. The reported activation energies were very close to the reference values obtained at 80% conversion for the four different coal/slag ratios, which is consistent with the way that the rate constant is usually calculated for a representative conversion interval (considering data points until conversion equal or higher than 80%).

Table 5: Activation energy and frequency factor calculated from Eqs. (8) and (12). Kinetic parameters reported data by Li et al. (2013) using the Avrami–Erofeev ($m = 2$) kinetic model. Temperature range from 1223 K to 1423 K

	EA (kJ/mol)			ko (min⁻¹)		
	Eq. (8)		A–E model	Eq. (12)		A–E model
	$X = 0.8$	$X = 0.5$		$X = 0.8$	$X = 0.5$	
Coal/slag ratio 1:0	102	76	112	1612	265	4806
Coal/slag ratio 1:1	91	71	94	1033	244	1625
Coal/slag ratio 1:2	84	77	87	662	501	965
Coal/slag ratio 1:3	56	53	53	61	58	52

The linear trend presented in Fig. 2 indicates that the reaction mechanism did not change in the studied temperature range. In all cases, the activation energies at 50% conversion were lower than those at 80% conversion. When the coal/slag ratio was 1:3, both values were almost the same, probably due to the slag acting as a catalyst, which can be associated with the alkali content. Results from this particular case led to the same conclusions as those obtained from the Silbermann et al. (2013) data in the previous section, i.e. the loss of catalyst as the gasification progresses. In this particular case, thermal annealing (Senneca et al., 1997) might not very well explain why the activation energy estimated at different conversions remains practically constant with excess amount of catalyst.

There was definitely no coherent trend with the frequency factor, which can be attributed to the existence of a maximum rate in the reported data detected at 1.5 min after the gases are switched by Li et al. (2013). For the highest temperature (1423 K), the total residence time to achieve an 80% conversion was 3.4 min for the less reactive sample (coal/slag ratio of 1:0), indicating that the time to replace the gases significantly affected the reading of the conversion rate in a considerable portion of the conversion range. The selected kinetic model overestimated the frequency factor, but was in reasonable agreement for the activation energy.

Using the proposed approach, it is possible to conclude that the initial part of the gasification affects the results. This effect is more pronounced as the conversion rate increases; therefore, at higher reaction temperatures or greater catalyst contents, the parameters calculated from a kinetic model could be significantly different with respect to the free-model calculations.

CO_2 Gasification Kinetics from Char Produced from Indian Coal Samples

Data reported by Mandapati et al. (2012) for CO_2 gasification of chars produced from four Indian coals (chars A, B, C, and D) were correlated using Eq. (8). Fig. 3 shows the plots of the logarithm of time versus the reciprocal of temperature for two different sets of conversion data points. Similar to the previous case, the experimental procedure involved a change of the reaction gas at the reaction temperature. The temperature range of the reported experiment was different between samples; however, it was significantly higher than the range where the reaction is thermodynamically limited. A linear trend for all four chars could be observed with a coefficient of determination higher than 0.99.

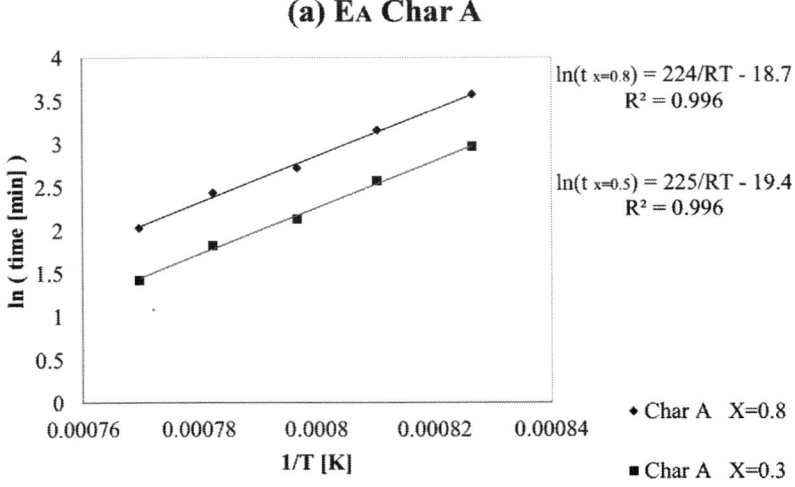

(a)

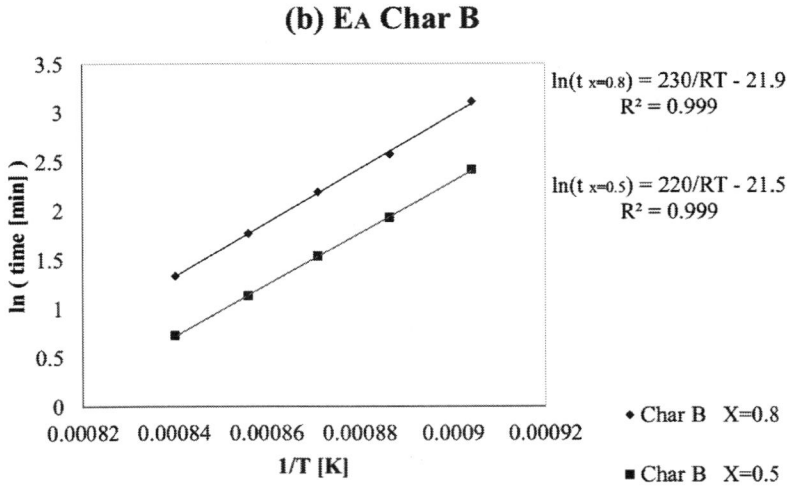

(b)

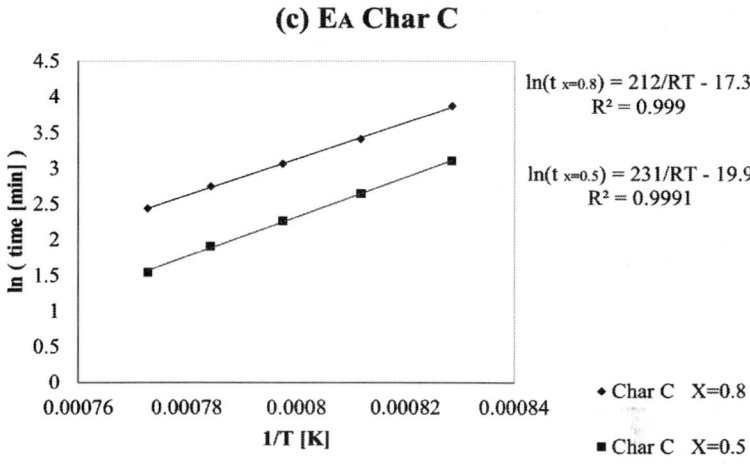

(c)

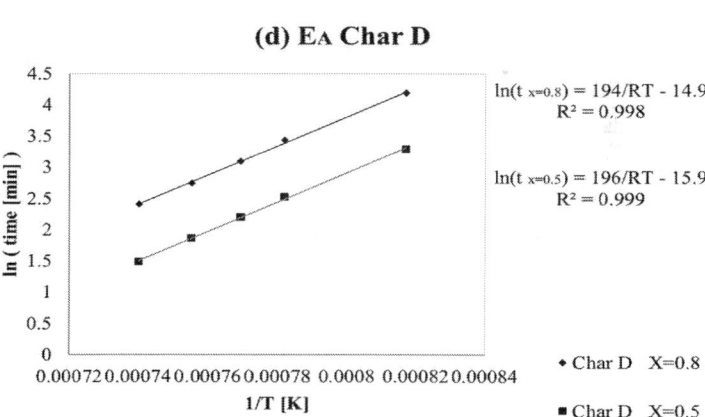

(d)

Figure 3: Logarithm of time versus reciprocal of temperature. Char CO_2 gasification: (a) Char A [937 °C to 1026 °C], (b) Char B [833 C to 917 °C]; (c) Char C [934 °C to 1021 °C], (d) Char D [951 C to 1078 °C]. Data from Mandapati et al. (2012).

Kinetic parameters are presented in Table 6, comparing the reference results obtained from Eqs.(8) and (12) for 80% and 50% conversions with the same reported parameters using the RPM model (columns 5 and 8, respectively). The reported activation energy values were in good agreement with the reference values proposed in this study. There was no significant difference between the reference activation energies calculated at 80% and 50% conversions, indicating that there was no loss of catalyst in these experiments and negligible mass transfer effects, which is consistent with the experimental procedure to reduce bed diffusion (Mandapati et al., 2012).

Table 6: Activation energy and frequency factor calculated from Eqs. (8) and (12). Kinetic parameters reported data by Mandapati et al. (2012) using the RPM

	Temperature range (K)	E_A (kJ/mol)			ko (min^{-1})		
		Eq. (8)		RPM	Eq. (12)		RPM
		X = 0.8	X = 0.5		X = 0.8	X = 0.5	
Char A	1210–1299	224	225	229	1.4E+08	2.7E+08	4.2E+09
Char B	1106–1190	230	220	213	3.4E+09	2.2E+09	1.1E+08
Char C	1207–1294	212	231	215	3.1E+07	4.4E+08	1.3E+08
Char D	1224–1351	194	196	193	2.9E+06	8.3E+06	1.8E+07

The time to observe a maximum rate after switching the gasifying agent was not reported; however, this time did not significantly affect the kinetics, since the time to achieve an 80% conversion for the less reactive char was almost three times higher than the previous study case (11.1 min for char D at 1351 K). The previous statement was verified, when parameter ψ (which should be higher than 2 if there really was a maximum rate as presented by Silbermann et al., 2013) of the RPM was checked in the original work:$\psi_A = 3.74$, $\psi_B = 3.19$, $\psi_C = 0.91$ and $\psi_D = 0.15$. This suggests that the model was considered nonlinear, but that the contribution of the nonlinear term was insignificant and just improved the determination coefficient of the regression.

The reported frequency factors were different with respect to the ones estimated from the free-model method, but with a similar magnitude order. Thus, the RPM with a ψ parameter close to 2 worked well and provided reasonable kinetic parameters, because those first minutes of the gasification (gas replacement into the reactor) that induced a maximum reaction rate were just a small part of the total residence time. If the experimental procedure was changed to avoid this situation, simpler models would give better correlations (Gomez et al., 2014 and Silbermann et al., 2013).

CO_2 Gasification Kinetics from Coal Plus Catalyst (K_2CO_3)

A recent study presenting the gasification of raw coal, ash-free coal, and ash-free coal plus catalyst (catalytic CO_2 gasification) was presented by Kopyscinski et al. (2013). The authors also presented a variation of the RPM called the extended random pore model (eRPM). Fig. 4 shows the plots of the logarithm of time versus the reciprocal of temperature for different sets of conversion data points.

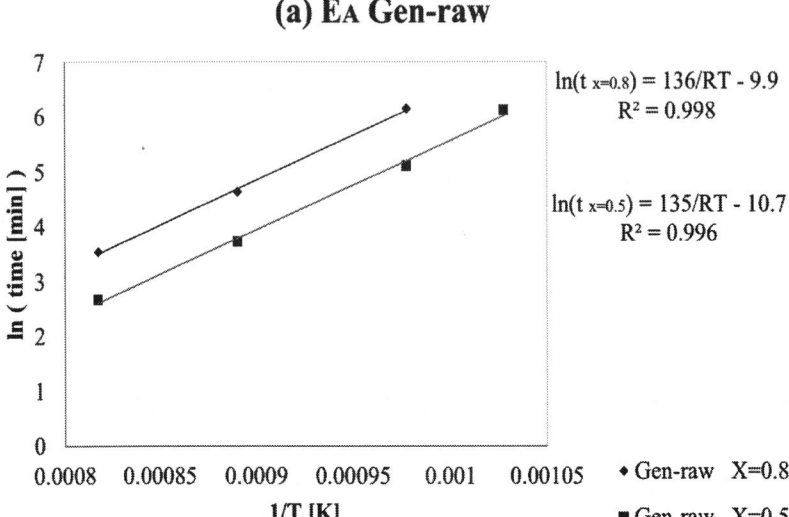

(a)

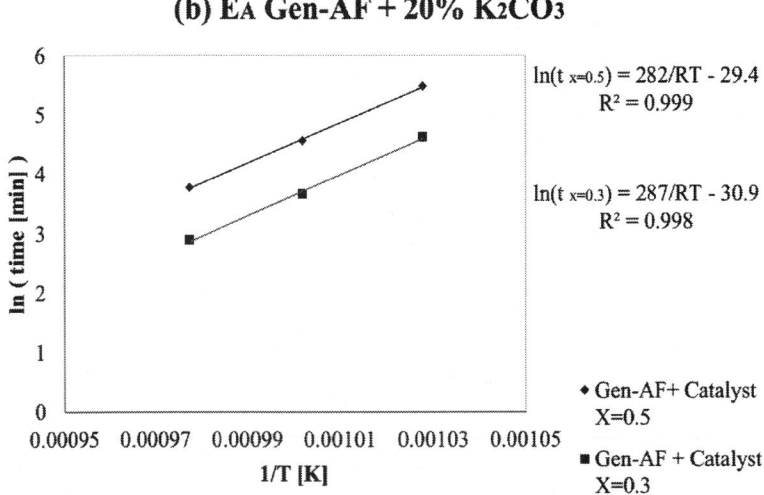

(b)

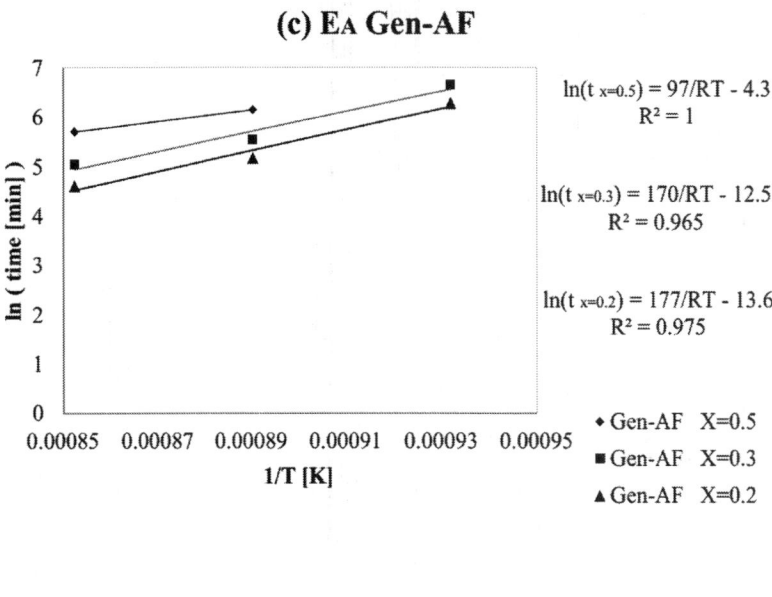

(c) E$_A$ Gen-AF

$$\ln(t_{x=0.5}) = 97/RT - 4.3$$
$$R^2 = 1$$

$$\ln(t_{x=0.3}) = 170/RT - 12.5$$
$$R^2 = 0.965$$

$$\ln(t_{x=0.2}) = 177/RT - 13.6$$
$$R^2 = 0.975$$

◆ Gen-AF X=0.5
■ Gen-AF X=0.3
▲ Gen-AF X=0.2

(c)

Figure 4: Logarithm of time versus reciprocal of temperature. Coal CO_2 gasification: (a) Gen-raw [700 C to 950 °C], (b) Gen-ash-free plus 20% K_2CO_3 [650 °C to 750 °C]; (c) Gen-ash-free [800 °C to 900 °C]. Data from Kopyscinski et al. (2013).

Linear trends could be observed, but the reported activation energies were quite different to the references obtained from Eq. (8). For the ash-free coal, an increase in temperature increased the activation energy, as presented in Fig. 4C; and, the reported data underestimated the real activation energy that could not be considered constant in the temperature range. With no alkali in the coal composition, the reactivity significantly decreased, and temperature must be increased for a practical constant reading of activation energy.

Complete information on the kinetic parameters obtained by the proposed method and the reported parameters from the RPM (ash-free coal) and eRPM (raw coal and ash-free plus catalyst) are presented in Table 7, including the temperature ranges and the reference temperature to calculate the rate constant. The activation energies for the proposed method and those reported for the three samples were similar. Even the activation energy values between the RPM and the eRPM were almost identical for the same coal type (Kopyscinski et al., 2013). The reason is similar to the study case of Mandapati et al. (2012) and related to the parameters of the RPM. Kopyscinski et al. (2013) reported the values of the parameter ψ for the RPM as $\psi_{Gen} = 0$, $\psi_{Gen\text{-}ash\text{-}free} = 0.45$ and $\psi_{Gen\text{-}ash\text{-}free+catalyst} = 0.0$ and those of the eRPM as $\psi_{Gen} = 4.3$ and $\psi_{Gen\text{-}ash\text{-}free+catalyst} = 64$.

Table 7: Activation energy and frequency factor calculated from Eqs. (8) and (12). Kinetic parameters reported data by Kopyscinski et al. (2013) using the RPM and eRPM

	Tempera-ture range (K)	EA (kJ/mol)		RPM	kT_{ref} (min⁻¹)		eRPM	Ref. tem-perature (K)
		Eq. (8)		RPM	Eq. (4)		eRPM	
		$X = 0.8$	$X = 0.5$		$X = 0.8$	$X = 0.5$		
Gen-raw	1023–1223	136	135	131	2.2E−03	5.5E−03	7.5E−04	1023
Gen-AF 20 wt% K_2CO_3	923–1023	282	287	264	4.2E−03	1.0E−02	1.1E−04	973
	Tempera-ture range (k)	EA (kJ/mol)		RPM	kT_{ref} (min⁻¹)		RPM	Ref. tem-perature (K)
		Eq. (8)		RPM	Eq. (4)		RPM	
		$X = 0.3$	$X = 0.2$		$X = 0.3$	$X = 0.2$		
Gen-AF	1073–1173	170	177	124	1.4E−03	2.0E−03	6.5E−04	1073

If parameter   was close to 2, the nonlinear part did not significantly affect the reaction rate. A large change of this parameter from the RPM to its modified version (eRPM) was mainly a consequence of an increase in the regression parameters. If the experimental procedure is performed without changing the gases, there will not be a maximum rate; and, the modeling can be reduced to a first-order reaction model.

A strange result was reported for the catalytic gasification of ash-free coal, since the activation energy was higher than the ones presented for the gasification of the ash-free and raw coals. Different explanations were given by Kopyscinski et al. (2013); however, there is an important fact that it was not considered: at the temperature range from 650 °C to 750 °C, Boudouard reaction is thermodynamically limited, and a comparison of activation energies at different temperature ranges is not appropriate. If the experiments for this coal type are performed at the same temperature range as the others, i.e. between 800 °C and 900 °C, a different result would probably be observed (i.e. lower activation energy) and the assumption of similar reaction mechanisms would make more sense.

Rate constants were presented instead of frequency factors, since Kopyscinski et al. (2013) reported them for different reference temperatures. The results were inconsistent for the ash-free coal plus catalyst compared with the other coal types: for example, using the eRPM at 1000 K as reference temperature, $k_{Gen-raw}$ was greater than $k_{Gen-ash-free+catalyst}$; however, the fastest coal at this temperature was the one with catalyst. This is a contradiction when Eq. (9) is considered (using the eRPM when the conversion is zero) and can be explained by the activation energy being obtained from a different temperature range where the reaction mechanism was different (Boudouard reactions just advance after 700 °C).

Steam Gasification Kinetics from Char Pyrolyzed at Different Pressures

The last experimental study analyzed in this work corresponds to steam gasification using data reported byFermoso et al. (2011) for four char samples (D1-1, D1-20, HI-1, HI-20), which were prepared from two different raw coals (DI, HV) and two different operation pressures during the pyrolysis (1 atm and 20 atm) at1000 °C. Fig.

5 shows the plots of the logarithm of time versus the reciprocal of temperature for two different sets of conversion data points. The experimental procedure involved the change of the reaction gas at the reaction temperature. The temperature range of the reported experiment was between 1173 K and 1323 K for all the experiments.

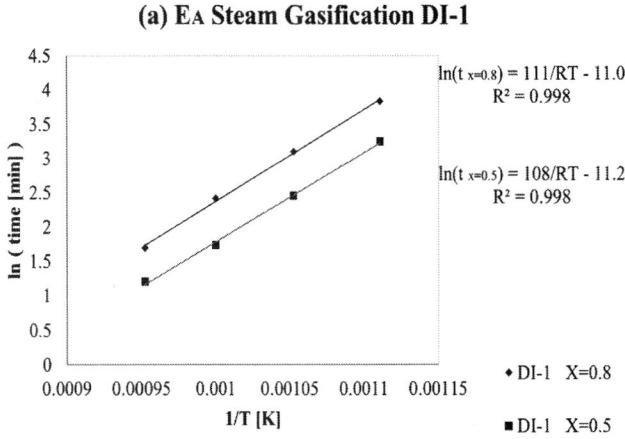

(a)

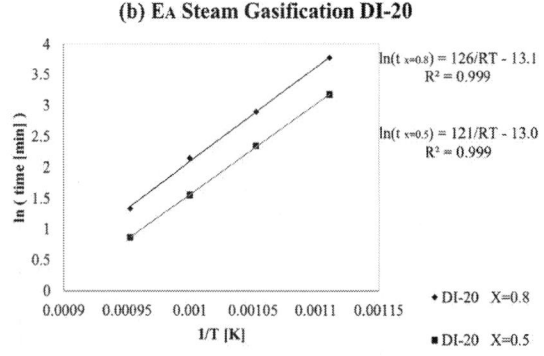

(b)

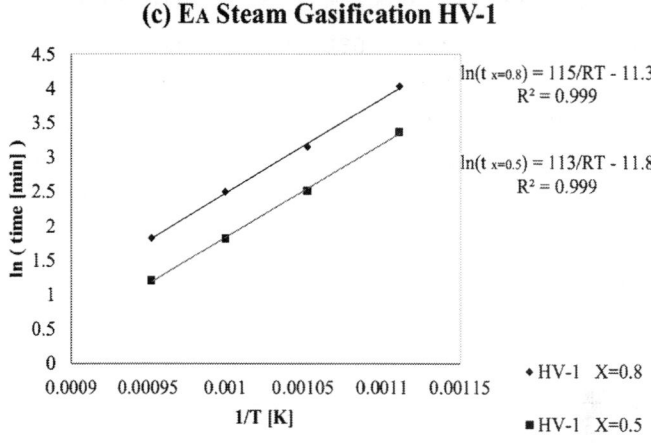

(c)

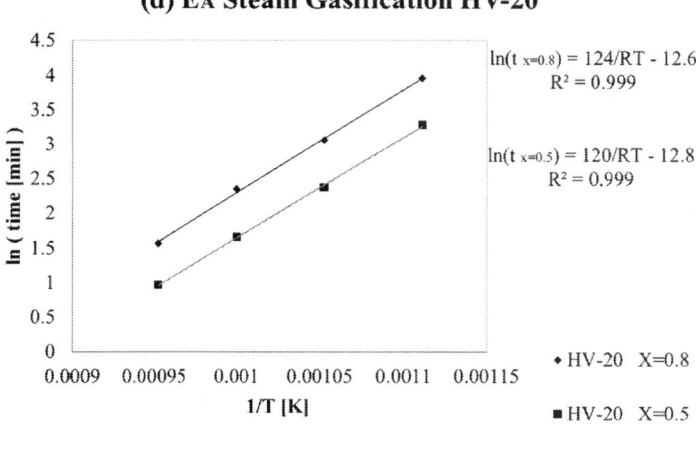

(d)

Figure 5: Logarithm of time versus reciprocal of temperature. Char steam gasification [30% vol H_2O–70% vol N_2] between 900 °C and 1050 °C: (a) DI-1, (b) DI-20; (c) HV-1, (d) HV-20. Data from Fermoso et al. (2011).

The RPM was selected by Fermoso et al. (2011) as the best kinetic model to fit the experimental data. The reported activation energies and frequency factors are presented in Table 8. These reference kinetic parameters were determined using Eqs. (8) and (10) for 80% and 50% conversions, respectively. Comparisons of the frequency factors and activation energies indicate that the reported data were overestimated, and the differences were significant.

Table 8: Activation energy and frequency factor calculated from Eqs. (8) and (12). Kinetic parameters reported data by Fermoso et al. (2011) using the RPM. Temperature range from 1173 K to 1323 K

	EA (kJ/mol)			k_o (min^{-1})		
	Eq. (8)		RPM	Eq. (12)		RPM
	$X = 0.8$	$X = 0.5$		$X = 0.8$	$X = 0.5$	
DI-1	111	108	178	5.9E+04	7.1E+04	1.6E+06
DI-20	126	121	200	4.8E+05	4.3E+05	1.6E+07
HV-1	115	113	183	8.4E+04	1.3E+05	2.5E+06
HV-20	124	120	195	3.0E+05	3.7E+05	7.8E+06

Activation energy must be considered carefully, since the reported data for steam gasification is close to 200 kJ/mol, which is an average value for CO_2 gasification at the same temperature range (Di Blasi, 2009,Irfan et al., 2011 and Silbermann et al., 2013). Analyzing other authors' works (Di Blasi, 2009 and Ren et al., 2013) in similar temperature ranges, the activation energies of steam gasification were significantly smaller than those of CO_2 gasification, which is in good agreement with the results of the proposed method. This can be attributed to the selection of the RPM as the best kinetic model, due to the gas switching performed during the experimental procedure.

There was no reported information about the time when the maximum rate was observed, but it is usually between one and two minutes with TGA (Gomez et al., 2014). The time to reach an 80% conversion at the highest temperature for all char samples was between 4 and 8 min. A similar reasoning to the previous study cases can be presented for the analysis of the value of parameter in the original reference. This maximum significantly affected the determination of the reaction rate and it is the reason why kinetic parameters cannot be calculated with this model for the particular reaction conditions.

Comparisons of the reference activation energies and rate constants at 50% and 80% conversions indicate that they were practically the same. This was a consequence of the higher reactivity of the char with steam than with CO_2: a faster reaction rate increased the accuracy of the proposed method. On the other hand, an increase in reactivity by increasing the temperature or adding catalyst decreased the accuracy of the RPM in determining the parameters of the Arrhenius equation, which is applicable if the experimental procedure induces a maximum reaction rate. If there is no maximum rate, there is no need for the RPM or other complex kinetic models.

CONCLUSIONS

A new method to obtain the parameters of the Arrhenius equation, i.e. activation energy and frequency factor, independent of the kinetic model has been presented and evaluated for five independent experimental studies. The plots of the logarithm of residence time, ln(*time*), for a particular conversion versus the reciprocal of temperature, $1/T$, followed a linear trend in the same way as an Arrhenius plot with a better coefficient of determination.

The slope of ln(*time*) versus $1/T$ yields the ratio of the activation energy to the ideal gas law constant (EA/R). The advantage of this method is that just one regression is required, rather than the $m + 1$ (m is the number of experimental temperatures) needed when a kinetic model is used. The uncertainty of the method is smaller, and it is less sensitive to any particular variation of the reaction rate, e.g.

when a maximum rate is induced due to the switching of the gases during steam and CO_2 gasification kinetic studies.

Activation energies and frequency factors calculated with the proposed method with Eqs. (8) and (12), respectively, produced consistent results. Their accuracy increased at high temperature ranges. These values can be used to test one single overall step kinetic models, which is a tool to scale-up industrial processes or validate assumptions about the reaction mechanism. Similar activation energies in two different range intervals indicated that the reaction mechanism was the same. Comparison of the estimated activation energy at different conversions (but the same range temperature) lets to prove assumptions about the constant amount of active catalyst and mass transfer limitations during the reaction progress.

Analyses of reported literature can be adjusted and kinetic parameters can be correctly compared, as was presented for five gasification experimental works: four CO_2 gasification studies and one steam gasification investigation. In particular, the random pore model did not estimate accurately the kinetic parameters of gasification, confirming the incidence of the reaction medium change generated by the experimental procedure.

ACKNOWLEDGMENTS

The authors are thankful to the Natural Sciences and Engineering Research Council (NSERC) of Canada for funding this study.

REFERENCES

1.   Ahmed, I.I., Gupta, A.K., 2011. Kinetics of woodchips char gasification with steam and carbon dioxide. Appl. Energy 88, 1613–1619.

2.   Bhatia, S.K., Perlmutter, D.D., 1980. A random pore model for fluid–solid reactions: I. Isothermal and kinetic control. AIChE J. 26, 379–386.

3.   Bhatia, S.K., Vartak, B.J., 1996. Reaction of microporous solids: the discrete random pore model. Carbon 34, 1383–1391.

4.   De Micco, G., Nasjleti, A., Bohe, A.E., 2012. Kinetics of the gasification of a Rio Turbio coal under different pyrolysis temperatures. Fuel 95, 537–543.

5.   Di Blasi, C., 2009. Combustion and gasification rates of lignocellulosic chars. Prog. Energy Combust. Sci. 35, 121–140.

6.   Duman, G., Uddin, Md.A., Yanik, J., 2014. The effect of char properties on gasification reactivity. Fuel Process. Technol. 118, 75–81.

7.   Fermoso, J., Gil, M.V., García, S., Pevida, C., Pis, J.J., Rubiera, F., 2011. Kinetic parameters and reactivity for the steam gasification of coal chars obtained under different pyrolysis temperatures and pressures. Energy Fuels 25, 3574–3580.

8.   Gomez, A., Silbermann, R., Mahinpey, N., 2014. A comprehensive experimental procedure for CO_2 coal gasification: is there really a maximum reaction rate? Appl. Energy 124, 73–81.

9.   Guizani, C., Escudero Sanz, F.J., Salvador, S., 2013. The gasification reactivity of high-heating-rate chars in single and mixed atmospheres of H_2O and CO_2. Fuel 108, 812–823.

10.   Hobbs, M.L., Radulovic, P.T., Smoot, L.D., 1993. Combustion and gasification of coals in fixed-beds. Prog. Energy Combust. Sci. 19, 505–586.

11.   Irfan, M., Usman, M.R., Kusakabe, K., 2011. Coal gasification in CO_2 atmosphere and its kinetics since 1948. A brief review. Energy 36, 12–40.

12.   Jeong, H.J., Park, S.S., Hwang, J., 2014. Co-gasification of coal–biomass blended char with CO_2 at temperatures of 900–1100 °C. Fuel 116, 465–470.

13.   Kim, H.S., Kudo Sh Tahara, K., Hachiyama, Y., Yang, H., Norinaga, K., et al., 2013. Detailed kinetic analysis and modeling of steam gasification of char from Ca-loaded lignite.

Energy Fuels 27, 6617–6631.

14. Kopyscinski, J., Habibi, R., Mims, Ch.A., Hill, J.M., 2013. K2CO3-catalyzed CO2 gasification of ash-free coal: kinetic study. Energy Fuels 27, 4875–4883.

15. Li, P., Yu, Q., Xie, H., Qin, Q., Wang, K., 2013. CO2 gasification rate analysis of Datong coal using slag granules as heat carrier for heat recovery from blast furnace slag by using a chemical reaction. Energy Fuels 27, 4810–4817.

16. Lin, L., Strand, M., 2014. Online investigation of steam gasification kinetics of biomass chars up to high temperatures. Energy Fuels 28, 607–613.

17. Loewenberg, D.A., Levendis, W.A., 1991. Combustion behavior and kinetics of synthetic and coal-derived chars: comparison of theory and experiment. Combust. Flame 84, 47–65.

18. Mandapati, R.M., Daggupati, S., Mahajani, S.M., Aghalayam, P., Sapru, R.K., Sharma, R.K., et al., 2012. Experiments and kinetic modeling for CO_2 gasification of Indian coal chars in the context of underground coal gasification. Ind. Eng. Chem. Res. 51, 15041–15052.

19. Nipattummakul, N., Ahmed, I., Kerdsuwan, S., Gupta, A.K., 2010. High temperature steam gasification of wastewater sludge. Appl. Energy 87, 3729–3734.

20. Popa, T., Fan, M., Argyle, M.D., Slimane, R.B., Bell, D.A., Towler, B.F., 2013. Catalytic gasification of a Powder River Basin coal. Fuel 103, 161–170.

21. Prabowo, B., Umeki, K., Yan, M., Nakamura, M.R., Castaldi, M.J., Yoshikawa, K., 2014. CO_2–steam mixture for direct and indirect gasification of rice straw in a downdraft gasifier: laboratory-scale experiments and performance prediction. Appl. Energy 113, 670–679.

22. Ren, L., Yang, J., Gao, F., Yan, J., 2013. Laboratory study on gasification reactivity of coals and petcokes in CO_2/steam at high temperatures. Energy Fuels 27, 5054–5068.

23. Senneca, O., Russo, P., Salatino, P., Masi, S., 1997. The

relevance of thermal annealing to the evolution of coal char gasification reactivity. Carbon 35, 141–151.

24. Silbermann, R., Gomez, A., Gates, I., Mahinpey, N., 2013. Kinetic studies of a novel CO2 gasification method using coal from deep unmineable seams. Ind. Eng. Chem. Res. 52, 14787–14797.

25. Singer, S.L., Ghoniem, A.F., 2011. An adaptive random pore model for multimodal pore structure evolution with application to char gasification. Energy Fuels 25, 1423–1437.

26. Su, J.L., Perlmutter, D.D., 1985. Effect of pore structure on char oxidation kinetics. AIChE J. 31, 973–981.

27. Umemoto, S., Kajitani, S., Hara, S., 2013. Modeling of coal char gasification in coexistence of CO_2 and H_2O considering sharing of active sites. Fuel 103, 14–21.

28. Wang, L., Sandquist, J., Varhegyi, G., Matas Güell, B., 2013. CO_2 gasification of chars prepared from wood and forest residue: a kinetic study. Energy Fuels 27, 6098–6107.

29. Woodruff, R.B., Weimer, A.W., 2013. A novel technique for measuring the kinetics of high-temperature gasification of biomass char with steam. Fuel 103,749–757.

30. Zhang, R., Wang, Q.H., Luo, Z.Y., Fang, M.X., Cen, K.F., 2014. Coal char gasification in the mixture of H_2O, CO_2, H_2, and CO under pressured conditions. Energy Fuels 28, 832–839.

CFD–DEM Simulation of Biomass Gasification with Steam in a Fluidized Bed Reactor

Xiaoke Ku, Tian Li, and Terese Løvås

Department of Energy and Process Engineering, Norwegian University of Science and Technology (NTNU), 7491 Trondheim, Norway

ABSTRACT

A comprehensive CFD–DEM numerical model has been developed to simulate the biomass gasification process in a fluidized bed reactor. The methodology is based on an Eulerian–Lagrangian concept, which uses an Eulerian method for gas phase and a discrete element method (DEM) for particle phase. Each particle is individually tracked and associated with multiple physical (size, density, composition, and temperature) and thermo-chemical (reactive or inert) properties. Particle collisions, hydrodynamics of dense gas-particle flow in fluidized beds, turbulence, heat and mass transfer, radiation, particle shrinkage, pyrolysis, and homogeneous and heterogeneous chemical reactions are all considered during biomass gasification with steam. A sensitivity analysis is performed to test the integrated model's response to variations in three different operating parameters (reactor temperature, steam/biomass mass ratio, and biomass injection position). Simulation results are analyzed both qualitatively and quantitatively in terms of particle flow pattern, particle mixing and entrainment, bed pressure drop, product gas composition, and carbon conversion. Results show that higher temperatures are favorable for the products in endothermic reactions (e.g. H_2 and CO). With the increase of steam/biomass mass ratio, H_2 and CO_2 concentrations increase while CO concentration decreases. The carbon conversion decreases as the height of injection point increases owing to both an increase of solid entrainment and a decrease of particle residence time and particle temperature. Meanwhile, the calculated results compare well with the experimental data available in the literature. This indicates that the proposed CFD–DEM model and simulations are successful and it can play an important role in the multi-scale modeling of biomass gasification or combustion in fluidized bed reactor.

INTRODUCTION

Due to the limited supply of conventional fossil fuels and global environmental problems, more and more attention has been paid to the renewable and clean energy technologies, among which biomass gasification is one of the most promising technologies for the efficient utilization of biomass. Biomass gasification is a complex thermo-chemical process in which biomass is converted into synthetic gas (syngas), a combination of hydrogen, carbon monoxide, and methane. The syngas could be then used as a fuel in internal combustion engines, gas turbines, or fuel cells for the production of heat, mechanical energy, or power, or as a feedstock for the synthesis of liquid fuels and chemicals. The fundamental aspects of biomass gasification have been mainly studied by experiments using lab-scale reactors (Gil et al., 1999, Qin et al., 2012 and Warnecke, 2000). Among the various gasification reactors, the fluidized bed (FB) reactor presents good prospects due to its high rates of heat and mass transfer, good temperature control, and its excellent mixing properties (Kern et al., 2013, Li et al., 2004 and Shen et al., 2008). In a typical FB reactor, fuel feed, together with inert bed material (e.g. sand) which acts as heat capacitance for the fuel, are fluidized by the gasifying agents, such as air (Kim et al., 2013), steam (Song et al., 2012), pure oxygen or their combination (Meng et al., 2011). There are many physico-chemical processes within a real biomass FB reactor, such as mixing, segregation, collision, particle heat-up, drying, pyrolysis, volatile matter combustion, and char reaction with O_2/steam/CO_2. Moreover their scales are greatly separated, which results in detailed study of the entire gasification process being a challenging task.

Computational fluid dynamic (CFD) models have become more and more popular in recognizing the dense gas–solid flow dynamics (Lathouwers and Bellan, 2001, Papadikis et al., 2010 and Ku et al., 2013) and chemical reactions (Ergüdenler et al., 1997, Nikoo and Mahinpey, 2008 and Sadaka et al., 2002) in FB reactors. Generally, all the CFD models developed can be broadly categorized into Eulerian–Eulerian and Eulerian–Lagrangian approaches. For Eulerian–Eulerian approach, both particle and fluid phases are treated as interpenetrating continua. It can predict the macroscopic characteristics of a system with relatively low computational cost and has actually dominated the modeling of fluidization process for many years (Gerber et al., 2010, Taghipour et al., 2005 and Wang et al., 2009). However, in addition to the difficulty of providing closure models for interaction terms between phases within its continuum framework, Eulerian–Eulerian approach does not recognize the discrete character of the particle phase and thus has trouble in modeling flows with a distribution of particle types and sizes. These difficulties can be naturally overcome by Eulerian–Lagrangian approach (Snider et al., 2011 and Xie et al., 2013) in which the gas is treated as continuous and particle as discrete phase. When the particle phase is solved by discrete element method (DEM), the

Eulerian–Lagrangian approach is also called CFD–DEM model. For CFD–DEM model, each particle is individually tracked and can be composed of multiple physical (size, density, composition, and temperature) and thermo-chemical (reactive or inert) properties. It can also offer detailed microscopic information at the particle level, such as particle trajectory, particle–particle and particle–fluid interaction, and transient forces acting on each particle, which is extremely difficult, even impossible to obtain by Eulerian–Eulerian approach. A crucial point when using CFD–DEM is the CPU load for particle collision monitoring as the number of particles increases. Thus, CFD–DEM simulations are often performed on the order of 10^4 particles and are mostly restricted to 2D or quasi-3D (domain width is one particle diameter) solutions. If chemical reactions are added, computation is more and more complicated and expensive. To date most of the CFD–DEM studies performed have been focused on the hydrodynamics of the isothermal fluidized bed and there have been few works on the simulation of dense gas–solid flow coupling with chemical reactions. Liu et al. (2011) used a CFD–DEM model to study char and propane combustion in a fluidized bed although their simulation conditions were strongly simplified, e.g., only 300 char particles were added at the start of simulation and there was no more fuel injection at later times. Bruchmüller et al. (2012) carried out a biomass fast pyrolysis simulation in a bubbling fluidized bed but did not take turbulence and chemical reactions into account. Gerber and Oevermann (2014) used a 2D CFD–DEM model to simulate wood gasification in a fluidized bed reactor but they used only charcoal as the bed material without any inert bed material such as sand used in ordinary experimental beds.

The aim of this study is to develop a comprehensive CFD–DEM model capable of describing dense, thermal, and *reactive* multi-phase flows like biomass gasification in a fluidized bed reactor. The model described here is an extension of our previous hydrodynamic CFD–DEM model. In our earlier paper (Ku et al., 2013), an isothermal and *non-reactive* CFD–DEM model was developed and applied to a series of test cases in order to quantify its predictive capabilities. These included (i) prediction of the characteristic fluidization behaviors (bubbles or slugs) of a typical bubbling fluidized bed, (ii) comparison of the minimum fluidization velocities predicted by different researchers, and (iii) comparison of the bed pressure drops generated by various drag correlations. The above comparisons performed have validated the hydrodynamic aspect of our CFD–DEM model. As a continuation, the hydrodynamic CFD–DEM model is enlarged here to account for the dense and reacting flows including models for turbulence, heat and mass transfer, radiation, particle shrinkage, pyrolysis, and heterogeneous and homogeneous reactions. The noteworthy novelties of the present CFD–DEM model include (i) a systematic presentation of the particle governing equations and gas transport equations within the Eulerian–Lagrangian concept, (ii) modeling of multiple homogeneous and heterogeneous reactions, (iii) resolving of turbulence by a k–ε model, (iv) 4×10^4 sand particles used as inert bed material and inter-particle and particle–wall collisions being resolved by a soft-sphere collision model, and (v) continuous biomass injection throughout the total simulation time. The integrated model is then applied to biomass gasification with steam in a lab-scale fluidized bed reactor. Simulation results are analyzed both qualitatively and quantitatively in terms of particle flow pattern, particle mixing and entrainment, bed pressure drop, composition distributions of product gas and other important characteristics in a fluidized bed reactor at different operating conditions (e.g. reactor temperature, steam/biomass mass ratio, biomass injection position). Besides, comparisons between calculated results and experimental data available in the literature are also carried out in order to verify the model.

This paper is organized as follows: In Section 2, the governing equations describing evolution of the particles and gas phase are firstly formulated. Herein, the sub-models of pyrolysis, char gasification, particle shrinkage, and gas phase reactions are also presented. In Section 3, the simulation setup is tabulated. In Section 4, the numerical results of biomass gasification with steam in a fluidized bed reactor are presented. Here, we first investigate the fluidization behavior, particle entrainment, and bed pressure drop. Then effects of different operating conditions, such as reactor temperature, steam/biomass mass ratio and biomass injection position, on the composition distributions of product gas and carbon conversion are documented where the CFD–DEM model is verified by comparing the calculated results with experimental data. Finally, a short summary and conclusions are given in Section 5. In addition, the symbols and subscripts used in the equations and abbreviations are described in the nomenclature at the end of the paper.

MATHEMATICAL MODELING

The CFD–DEM model is formulated based on an unsteady-state Eulerian–Lagrangian multiphase model meaning transport equations are solved for the continuous gas phase and each of discrete particles is tracked through the calculated gas field. The interaction between the continuous phase and the discrete phase is taken into account by treating the exchange of mass, momentum and energy between the two systems as source terms in the governing equations. Specifically, the mechanisms of mass and energy exchange are adopted from the work of Kumar and Ghoniem (2012) with certain modifications as will be outlined below. Furthermore, for momentum exchange, detailed implementation issues are available in our earlier publication (Ku et al., 2013).

Discrete Particle Phase

The discrete particle phase consists of sand and biomass particles which are modeled in a Lagrangian manner. Sand plays only the role of heat carrier in biomass gasification without taking part in any reactions, whereas biomass undergoes successive physical and chemical processes such as heat-up, drying, pyrolysis, and gasification and its behavior is strongly related to operating conditions.

Particle Motion

The governing mass, momentum, and energy equations for each particle are as follows,

Mass:

$$\frac{dm_p}{dt} = \frac{dm_{vapor}}{dt} + \frac{dm_{devol}}{dt} + \frac{dm_{C-CO_2}}{dt} + \frac{dm_{C-H_2O}}{dt} \tag{1}$$

Momentum:

$$m_p \frac{d\mathbf{v}_p}{dt} = \mathbf{f}_g + \mathbf{f}_c + m_p \mathbf{g} \tag{2}$$

$$I_p \frac{d\boldsymbol{\omega}_p}{dt} = \mathbf{T}_p \tag{3}$$

$$\mathbf{f}_g = \frac{V_p \beta}{\varepsilon_p}(\mathbf{u}_g - \mathbf{v}_p) \tag{4}$$

$$\beta = \begin{cases} 150\frac{\varepsilon_p^2 \mu_g}{\varepsilon_g^2 d_p^2} + 1.75\frac{\varepsilon_p \rho_g}{\varepsilon_g d_p}|\mathbf{u}_g - \mathbf{v}_p| & \varepsilon_g < 0.8 \\ \frac{3}{4}C_d \frac{\varepsilon_p \rho_g}{d_p}|\mathbf{u}_g - \mathbf{v}_p|\varepsilon_g^{-2.65} & \varepsilon_g \geq 0.8 \end{cases} \tag{5}$$

$$C_d = \begin{cases} \frac{24}{Re_p}(1 + 0.15Re_p^{0.687}) & Re_p < 1000 \\ 0.44 & Re_p \geq 1000 \end{cases} \tag{6}$$

$$Re_p = \varepsilon_g \rho_g d_p |\mathbf{u}_g - \mathbf{v}_p| / \mu_g \tag{7}$$

Energy:

$$m_p c_p \frac{dT_p}{dt} = hA_p(T_g - T_p) + \frac{\varepsilon_p A_p}{4}(G - 4\sigma T_p^4) + Q_p \tag{8}$$

As shown in Eq. (2), $\mathbf{f}_c$, i.e. the total contact force acting on particle due to inter-particle or particle–wall collisions, is taken into account and it is necessary for dense gas-particle flows. This is different from the model of Kumar and Ghoniem (2012) which does not consider the contact forces and thus their model is only applicable to dilute multiphase systems.

Here, the inter-phase momentum exchange coefficient β is modeled via the well-known Gidaspow drag correlation (Gidaspow, 1994). As shown in Eq. (5), the Gidaspow model combines Ergun (1952) and Wen and Yu (1966) correlations for the dilute and dense granular regime where a porosity ε_g of 0.8 is adopted as the boundary between these two regimes. This model is often used in the literature and effects of using different drag models were discussed in earlier publication (Ku et al., 2013).

As shown in Eq. (8), the particle temperature is calculated taking into account the heat transfer due to convection, radiation, and source term Q_p including both the latent heat of vaporization of water from the particle to the gas phase and the heat generated by the heterogeneous char reactions.

The inter-particle or particle–wall collisions are resolved by a soft-sphere discrete element method which was firstly proposed by Cundall and Strack (1979). In this method, the inter-particle contact forces are calculated using equivalent simple mechanical elements, such as spring, slider and dashpot (see Fig. 1). Particles are allowed to overlap slightly. The normal force tending to repulse the particles can then be deduced from this spatial overlap and the normal relative velocity at the contact point. The spring stiffness can be calculated by Hertzian contact theory when the physical properties such as Young's modulus and the Poisson ratio are known. A characteristic feature of the soft-sphere model is that it is capable of handling multiple particle–particle contacts which is of much importance when modeling dense particle systems like fluidized bed. Detailed implementation issues of the soft-sphere model are available in the literature (e.g. Tsuji et al., 1992), which are not stated here for the sake of shortness. In this study, the following physical properties are adopted for the collision model: Young's modulus is 5×10^6 Pa; Poisson ratio is 0.3; coefficient of restitution and friction coefficient are 0.9 and 0.3, respectively. All values are equally valid for walls and particles (Bruchmüller et al., 2012 and Ku et al., 2013).

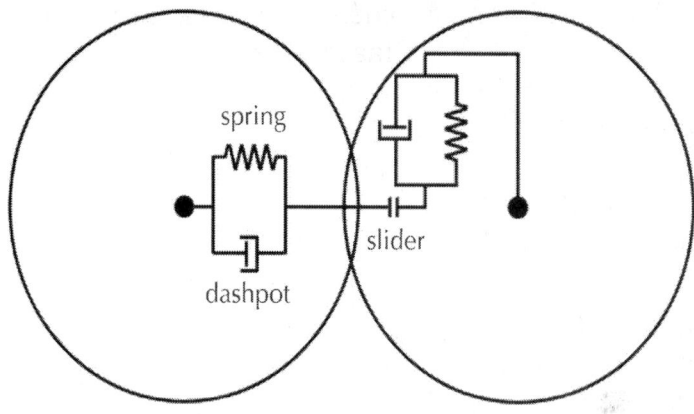

Figure 1: The spring–slider–dashpot collision model.

Pyrolysis

As soon as fresh biomass is fed into the bottom of the hot sand bed, it is immediately heated up, and thereby the devolatilization and pyrolysis of biomass as well as char gasification occurs. The pyrolysis compositions released from biomass can be expressed by the following equilibrium equation and each product yield is solved with the help of the elemental conservation analysis.

$$\text{Biomass} \rightarrow \alpha_1 CO + \alpha_2 H_2O + \alpha_3 CO_2 + \alpha_4 H_2 + \alpha_5 CH_4 + \alpha_6 \text{char(s)}$$
$$+ \alpha_7 \text{ash(s)}, \ \sum_i \alpha_i = 1 \tag{9}$$

Note that, in the present model, reactions with sulfur and nitrogen are not taken into account due to their little amount (see Table 3), and they are considered passing directly to ash. CH_4 is the only hydrocarbon species taken into consideration. Although C_2H_2, C_2H_4, C_2H_6, and other higher hydrocarbons (tar) are produced in the pyrolysis process, they are treated as non-stable products and this mechanism has also been widely used by other researchers (Ergüdenler et al., 1997 and Gerber et al., 2010).

Consistent with Abani and Ghoniem's work (Abani and Ghoniem, 2013), the devolatilization rate is modeled using a single step first-order Arrhenius reaction.

$$\frac{dm_{\text{devol}}}{dt} = -A\exp\left(-\frac{E}{RT_{\text{p}}}\right)m_{\text{devol}}$$

(10)

where m_{devol} is the mass of the volatiles remaining in the particle, $A=5.0\times10^6$ s^{-1}, and $E=1.2\times10^8$ J/kmol (Prakash and Karunanithi, 2008). The devolatilization process is assumed to be energetically neutral because the heat of devolatilization is generally negligible as compared to heat of reactions due to char consumption reactions (Abani and Ghoniem, 2013).

Char Conversion Chemistry

After devolatilization, the biomass particle is left with char and ash. Ash is assumed to be carried along with the particle without taking part in any reactions. Char will react in the presence of carbon dioxide and steam and gets converted into carbon monoxide and hydrogen. The following heterogeneous reactions are assumed and implemented in OpenFOAM.

$C+CO_2\rightarrow2CO$ (R1)

$C+H_2O\rightarrow CO+H_2$ (R2)

Reactions R1 and R2 are endothermic gasification reactions and R1 is known as the Boudouard reaction.

The char consumption rate which includes the effects of both diffusion and kinetic rates is given as

$$\frac{dm_{C-i}}{dt} = -A_p p_i \frac{r_{\text{diff},i} r_{\text{kin},i}}{r_{\text{diff},i} + r_{\text{kin},i}}$$

(11)

$$r_{\text{diff},i} = C_i \frac{[(T_p + T_g)/2]^{0.75}}{d_p}$$

(12)

$$r_{\text{kin},i} = A_i T_p \exp\left(\frac{-E_i}{RT_p}\right)$$

(13)

where m_{c_i} is the mass of the char remaining in the particle when char reacts with gasifying species $i(=CO_2,$ or $H_2O)$, pi is the partial pressure of the gasifying species, $r_{\text{diff},}i$ and $r_{\text{kin},}i$ are the diffusion rate and the kinetic rate, respectively. C_i is the mass diffusion rate constant. Ai and Ei are the parameters typical of the Arrhenius forms of kinetic rates. For wood biomass considered in the present study, the constants used for kinetic and diffusion rates are assembled below in Table 1 (Abani and Ghoniem, 2013).

Table 1: Heterogeneous reaction constants

Parameters	Values
A_{H2O} (s/(m K))	45.6
E_{H2O} (J/kmol)	4.37×10^7
A_{CO2} (s/(m K))	8.3
E_{CO2} (J/kmol)	4.37×10^7
Ci ($i=H_2O$, CO_2) (s/K$^{0.75}$)	5.0×10^{-12}

Particle Shrinkage

The char–gas chemistry consumes the solids and biomass particles shrink as they react with the gas phase. Particle shrinkage not only has an effect on gasification but also strongly affects particle trajectory on its way out of the reactor. Without particle shrinkage char entrainment will be highly over-predicted. Here we assume that particle density ($_p$) stays constant throughout the gasification process and a mass-proportional shrinkage is adopted for each biomass particle. Thus the diameter of biomass particle shrinks as follows (Bruchmüller et al., 2012),

$$d_{\mathrm{p}} = \left(\frac{6m_{\mathrm{p}}}{\pi\rho_{\mathrm{p}}} \right)^{1/3}$$

(14)

Continuous Gas Phase

The gas phase is modeled as a continuum, known as an Eulerian type model.

Gas Phase Motion

For continuum gas phase, the governing mass, momentum, energy, and species transport equations can be typically represented by the following equations.

Mass:

$$\frac{\partial}{\partial t}(\varepsilon_{\mathrm{g}}\rho_{\mathrm{g}}) + \nabla \cdot (\varepsilon_{\mathrm{g}}\rho_{\mathrm{g}}\mathbf{u}_{\mathrm{g}}) = S_{\mathrm{p,m}}$$

(15)

Momentum:

$$\frac{\partial}{\partial t}(\varepsilon_{\mathrm{g}}\rho_{\mathrm{g}}\mathbf{u}_{\mathrm{g}}) + \nabla \cdot (\varepsilon_{\mathrm{g}}\rho_{\mathrm{g}}\mathbf{u}_{\mathrm{g}}\mathbf{u}_{\mathrm{g}}) = -\nabla p + \nabla \cdot (\varepsilon_{\mathrm{g}}\boldsymbol{\tau}_{\mathrm{eff}}) + \varepsilon_{\mathrm{g}}\rho_{\mathrm{g}}\mathbf{g} + S_{\mathrm{p,\ mom}}$$

(16)

Energy:

$$\frac{\partial}{\partial t}(\varepsilon_{\mathrm{g}}\rho_{\mathrm{g}}E) + \nabla \cdot (\varepsilon_{\mathrm{g}}\mathbf{u}_{\mathrm{g}}(\rho_{\mathrm{g}}E + p)) = \nabla \cdot (\varepsilon_{\mathrm{g}}\alpha_{\mathrm{eff}}\nabla h_{\mathrm{s}}) + S_h + S_{\mathrm{p,h}} + S_{\mathrm{rad}}$$

(17)

$$E = h_s - \frac{p}{\rho_g} + \frac{u_g^2}{2}$$
(18)

Species:

$$\frac{\partial}{\partial t}(\varepsilon_g \rho_g Y_i) + \nabla \cdot (\varepsilon_g \rho_g \mathbf{u}_g Y_i) = \nabla \cdot (\varepsilon_g \rho_g D_{\text{eff}} \nabla Y_i) + S_{p,Y_i} + S_{Y_i}$$
(19)

Note that the above transport equations have taken the volume fraction of gas ε_g into account and are applicable to the dense and reactive gas–particle flow in fluidized beds studied in this paper. They are different from the ones of Kumar and Ghoniem (2012) which do not consider ε_g and are only suitable for very dilute gas–particle flows.

Here, the effective stress tensor, τ_{eff}, is the sum of the viscous and turbulent stresses. Similarly the effective dynamic thermal diffusivity α_{eff} and mass diffusion coefficient for species D_{eff} take both the viscous and turbulent contributions into account. P-1 radiation model is adopted to solve the radiation source term S_{rad} as it has generally been chosen in CFD simulations of pulverized fuel gasification with radiation scattering (Backreedy et al., 2006).

As shown by Eq. (19), a transport equation is solved for each gas species, and the total gas phase properties are calculated from the mass fractions of the gas species making up the gas mixture. The mass, momentum, and enthalpy Eqs. (15), (16) and (17), respectively, are solved at each time step for the gas mixture. The flow is compressible, and the gas phase pressure, volume, temperature, and density are related through equations of state.

In order to solve turbulence, the governing transport equations for k and ε, which take into account the volume fraction of gas ε_g and are suitable for our dense gas–particle simulation system, are as follows (Kumar and Ghoniem, 2012 and Wang et al., 2009),

$$\frac{\partial}{\partial t}(\varepsilon_g\rho_g k)+\nabla\cdot(\varepsilon_g\rho_g \mathbf{u}_g k)=\nabla\cdot\left(\varepsilon_g\left(\mu_g+\frac{\mu_t}{\sigma_k}\right)\nabla k\right)+\varepsilon_g G_k-\varepsilon_g\rho_g\varepsilon$$

(20)

$$\frac{\partial}{\partial t}(\varepsilon_g\rho_g\varepsilon)+\nabla\cdot(\varepsilon_g\rho_g \mathbf{u}_g\varepsilon)=\nabla\cdot\left(\varepsilon_g\left(\mu_g+\frac{\mu_t}{\sigma_\varepsilon}\right)\nabla\varepsilon\right)+\varepsilon_g\frac{\varepsilon}{k}(C_{\varepsilon 1}G_k-C_{\varepsilon 2}\rho_g\varepsilon)$$

(21)

The constants $C\varepsilon_1$=1.44, $C\varepsilon_2$=1.92, σk=1.0, and $\sigma\varepsilon$=1.3. The turbulent viscosity μ_t is computed as a function of k and ε,

$$\mu_t=\rho_g C_\mu\frac{k^2}{\varepsilon}$$

(22)

where $C\mu$ is a constant which is set as 0.09.

Gas Phase Reactions

There are hundreds of gas phase chemical reactions in a gasification reactor. Even if all the elemental reactions and their rates of reaction could be identified, it is not possible to calculate so large number of coupled reactions. For the sake of simplification, a reduced set of 2 global reactions (3 reactions considering reverse reaction) is used to describe the major conversion rates in the reactor and effect of turbulence on reactions is resolved by the partially stirred reactor (PaSR) model (Abani and Ghoniem, 2013). Chemical reaction equations and their reaction rates as well as adopted references are listed inTable 2. The reaction rate is in kmol/(m³ s), and [·] implies mole concentration (kmol/m³) of the gas species enclosed in the brackets. Reactions R3 is the consumption of CH_4 through steam reforming. Reaction R4 is known as the reversible water–gas shift reaction. Both forward reaction rate k_f and reverse reaction rate k_b of R4 are calculated in lieu of a combined forward–reverse rate and k_f and k_b are related by the equilibrium constant $k_{eq}=k_f/k_b$.

Table 2: Considered homogeneous chemical reactions and their reaction rates

Reactions		Reaction rate	Refs.
$CH_4+H_2O \rightarrow CO+3H_2$	(R3)	$k=3.0\times10^8[_{CH4}][_{H2}O]$ $\exp(-1.26\times10^8/RT)$	Jones and Lindstedt (1988)
$CO+H_2O \rightarrow CO_2+H_2$	(R4)	$_{kf}=2.78\times10^3[CO][_{H2}O]$ $\exp(-1.26\times10^7/RT)$ $_{kb}=9.59\times10^4[_{CO2}][_{H2}]$ $\exp(-4.66\times10^7/RT)$ $_{keq}=0.029\exp(3.40\times10^7/RT)$	Gómez-Barea and Leckner (2010)

Computational Methodology

Since the governing equations for particles and the gas phase are different, different solution schemes have to be used. For discrete particles, a first-order Euler time integration scheme is used to solve the translational and rotational motions of particles. Inter-particle and particle–wall collisions are modeled by soft-sphere collision method (see Fig. 1), where the solution scheme is well documented in the literature (e.g. Tsuji et al., 1992). Meanwhile, the drying, pyrolysis, and gasification submodels update particle properties like temperature, diameter, composition, and heat capacity at each fluid time step. For continuous gas phase, time discretization of the transporting equations is based on an Euler scheme and spatial discretization uses a finite-volume technique. The coupling between the discrete particles and the gas phase is achieved by the inter-phase source terms ($S_{p,m}$, $S_{p,mom}$, $S_{p,h}$, $S_{p,Yi}$), which are solved at every fluid time step. All mathematical models and schemes described above have been developed and implemented into an open source C++ toolbox OpenFOAM (OpenCFD Ltd, 2012). The codes are made parallel and each case shown in the following sections takes about 14 days running time on a 16-core Intel node to accomplish the 20 s real time of simulation.

SIMULATION SETUP

All calculations are performed on a lab-scale biomass fluidized bed reactor which is taken from the experimental study of Song et al. (2012). Fig. 2 shows a sketch of the simulated geometry. It consists of a rectangular container of dimensions 0.23 m (width)×1.5 m (height)×0.0015 m (thickness) with a orifice of 0.01 m in width at the center of the bottom wall. The left, right, bottom walls, the bottom orifice and the top outlet compose the whole calculation domain boundaries. Initially, the reactor is filled completely with N_2 and a packed sand bed which is composed of 40,000 spherical sand particles with a diameter of 1.5 mm. The initial temperature of the sand and the gas in the domain is set equal to the operating reactor temperature (T_r). Hence, although the sand bed is initially stationary, it is assumed that it has been preheated. At the bottom inlet, mass flow rates for gas and biomass are specified, respectively. At the walls, no-slip conditions are applied for the gas phase and the wall temperature is specified according to the operating reactor temperature. At the top outlet, the atmospheric pressure boundary condition is adopted and particles are allowed to exit the computational domain during the simulation, modeling a fine solids entrainment phenomenon.

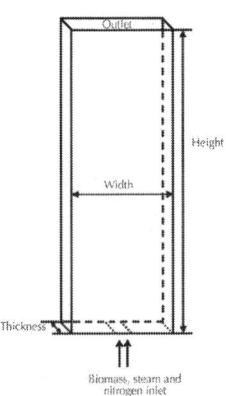

Figure 2: Geometry of the fluidized bed reactor.

In the simulations, biomass is fed through the bottom orifice, together with a mixture of steam and nitrogen which is used as the gasifying agent as well as the fluidizing gas. The initial diameter of biomass particle is 1.5 mm which is taken from the experiment. Pine wood is used as the biomass fuel and its initial properties, such as proximate and elemental analyses, are given in Table 3. The operating conditions such as reactor temperatures (T_r), biomass feed rate, and steam/biomass mass ratio (S/B), are in accordance with Song et al.'s (2012) experiment data. Table 4 summarizes the parameter settings used in the simulation and the boundary conditions for the gas phase are listed in Table 5. Note that all simulation cases are performed with a bottom biomass injection (see Fig. 2) except in Subsection 4.6. "Effect of biomass injection position" where the particle behaviors are compared among three different injection positions.

Table 3: Pine wood properties (Song et al., 2012)

Proximate analysis (wt%, on the as-received basis)		Elemental analysis (wt%, on the daf basis)	
Moisture	11.89	C	46.29
Ash	1.56	H	6.48
Volatile	71.78	O	46.08
Fixed carbon	14.77	N&S	1.15

Table 4: Parameter settings for the simulation system

Property	Value	Property	Value
Bed size, (m)	0.23×1.5×0.0015	Sand density, (kg/m³)	2600
Reactor temperature, (°C)	820, 870, 920	Sand specific heat, (J/ (kg K))	860
CFD cell size, (m)	0.01×0.02×0.0015	Sand number, (dimensionless)	40,000
Fluid time step, (s)	$1.0×10^{-5}$	Biomass type, (dimensionless)	pine

Total simulation time, (s)	20	Initial biomass diameter, (mm)	1.5
Particle shape, (dimensionless)	Spherical	Biomass density, (kg/m³)	470
Collision restitution coefficient, (dimensionless)	0.9	Biomass specific heat, (J/(kg K))	1500
Particle friction coefficient, (dimensionless)	0.3	Biomass feed rate, (g/s)	0.03125
Solid emissivity, (dimensionless)	0.9	Gas density, ρ_g	⊠
Sand diameter, (mm)	1.5	Inlet gas flow rate, (g/s)	0.18935

* $_g$ is determined based on the gas equation of state.

Table 5: Boundary conditions for gas phase in the simulation

Boundaries	Velocity	Pressure	Temperature	Porosity
Left and right walls	No slip	Zero gradient	Fixed value	Zero gradient
Bottom wall	No slip	Zero gradient	Zero gradient	Zero gradient
Inlet orifice (bottom)	Fixed flow rate	Zero gradient	Fixed value	Fixed value
Outlet (top)	Zero gradient	Fixed value	Zero gradient	Zero gradient

RESULTS AND DISCUSSIONS

Initial Bed Preparation

As described in Section 3, an initial packed sand bed is needed to start the fluidized bed simulation and it is generated as follows. The container is uniformly divided into a set of small rectangular lattices throughout the calculation domain. Then 40,000 sand particles with zero velocity are positioned at the centers of these lattices and allowed to fall down under the influence of gravity in the absence of inlet jet gas. As shown inFig. 3, pluvial deposition of the particles finally results in a static bed of height about 0.35 m and porosity around 0.42. This deposited bed is then used as the initial packed bed for the fluidized bed gasification simulation. As pointed out by Xu and Yu (1997), the initial input data for this deposited bed include not only the particle coordinates but also the forces and torques which come with the deposition of particles in the packing process.

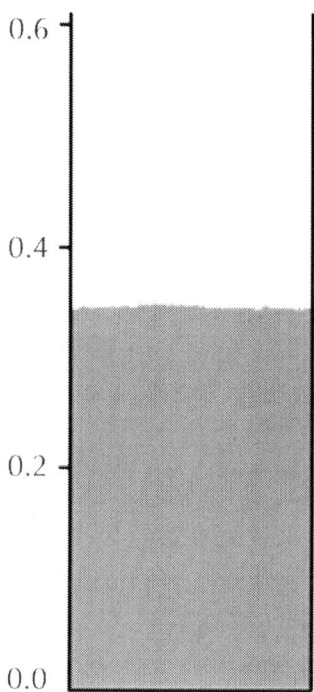

Figure 3: Particle configurations after a simulated packing process.

Fluidization Behavior

To investigate the fluidization behavior of the bed, the formation and development of bubbles with time are firstly illustrated. Fig. 4 shows the simulated particle flow patterns with the time increment being 0.1 s at the beginning of simulation, representative for the base case (T_r=820 °C, S/B=1.2). Particles are colored by solid type. Brown color indicates sand particle and black color denotes biomass. Overall, the conditions in the reactor are almost symmetrical at the beginning of simulation. As an initial response of the bed to the introduction of fluidizing gas, a significant upward flow of particles is caused due to the instantaneous breakup of the inter-particle locking. It is readily observed that a big bubble (void structure) with an oval shape is firstly formed at the jet region (t=0.1 s), which forces particles in its front to rise. This bubble grows as gas flows upward and eventually collapses (t=0.2 s, 0.3 s). At later times, new bubbles continue to form at the bottom of bed and then they undergo the same procedure. Besides the bubble formation, the existence of "slug" structure at the upper part of the bed is also clearly predicted (t=0.4 s). The term "slug" is used here to describe a dilute region of particles which occupies the whole width of the bed and a similar definition is also given by other investigators (Hoomans et al., 1996 and Kafui et al., 2002). The formation of bubbles and slugs in a typical fluidized bed reactor was also reported in the literature both numerically (Boyalakuntla, 2003, Hoomans et al., 1996 and Xu and Yu, 1997) and experimentally (Tsuji et al., 1993). Att=0.40 s, a bed expansion estimated at 120% of the initial bed height is observed. Fig. 4 also shows the biomass particles (in black color), which start to enter into the reactor at t=0 through the bottom orifice, move up inside the dense sand bed.

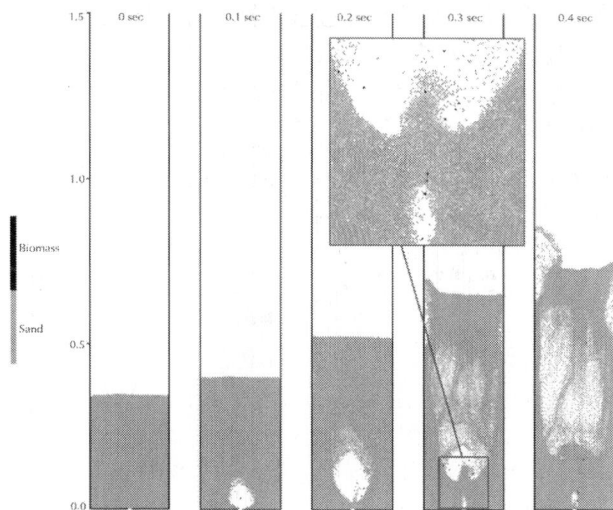

Figure 4: Particle flow patterns with the time increment being 0.1 s at the beginning of simulation. T_r=820 °C, S/B=1.2. (For interpretation of the references to color in this figure the reader is referred to the web version of this article).

Fig. 5 depicts the particle flow patterns with the time increment being 0.1 s at the end of simulation. Generally, due to the gas productions from biomass by pyrolysis and gasification, the conditions in the reactor are not symmetrical and the bed is in a churned-turbulent state. It is observed that the inlet jet degenerate into bubbles, which rise through the bed and grow by coalescence with other bubbles to form slugs. When bubbles and slugs burst at the bed surface, particles tend to be pushed towards the wall and then fall down along the wall. This provokes a quite vigorous fluidization and strong mixing takes place. It is easily seen that biomass particles are relatively evenly distributed throughout the dense sand bed, illustrating the effectiveness for particle mixing which is regarded as a special characteristic of fluidized beds. Good mixing favors the direct contact between virgin cold biomass and hot sand and in turn allows a good heat transfer.

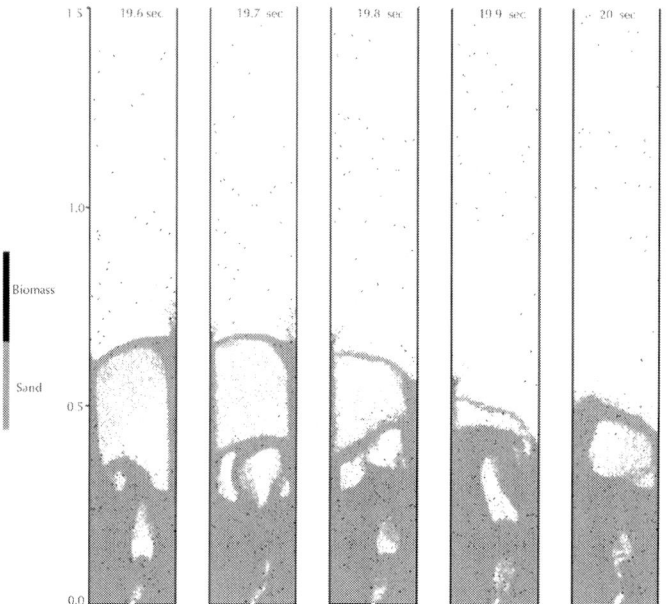

Figure 5: Particle flow patterns with the time increment being 0.1 s at the end of simulation. T_r=820 °C, S/B=1.2.

Fig. 6 depicts the snapshot of particle temperatures at the end of simulation. It is easily observed that the sand particles play the role of heat carrier and they have a temperature which is very close to the operating temperature (T_r=820 °C). At the same time, the strong mixing demonstrated in Fig. 5 favors the direct contact between virgin cold biomass and hot sand and results in a quick increase in the biomass temperature, whereas most of the biomass particles still have a relatively lower temperature compared to sand particles as shown in Fig. 6.

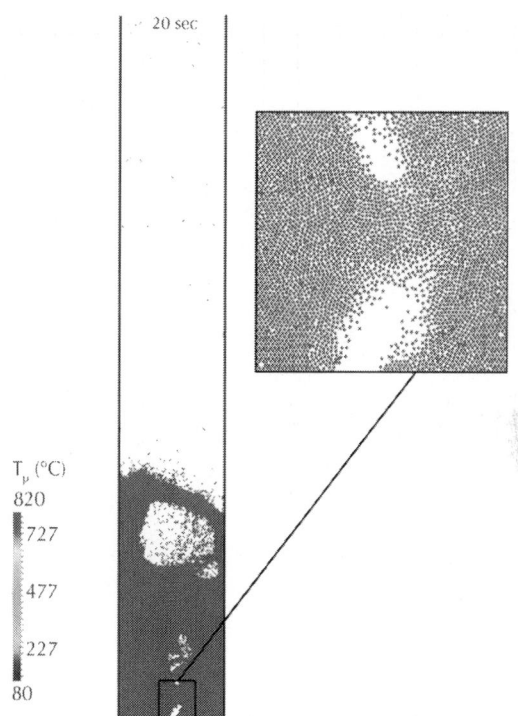

Figure 6: Snapshot of particle temperatures at the end of simulation. T_r=820 °C, S/B=1.2.

To show the transient behavior due to the fluidization of the bed, the pressure drop across the bed Δp is plotted in Fig. 7 as a function of time t. Δp is obtained as the difference between the average gas pressure in the bottom and top rows of the computational cells. It is easily observed that Δp fluctuates with time. The bed pressure drop fluctuations in a bubbling fluidized bed are considered to be caused by bubbles and slugs that form and collapse at regular intervals (Boyalakuntla, 2003) and effects of different drag models on the bed pressure drop has been discussed in our earlier paper (Ku et al., 2013).

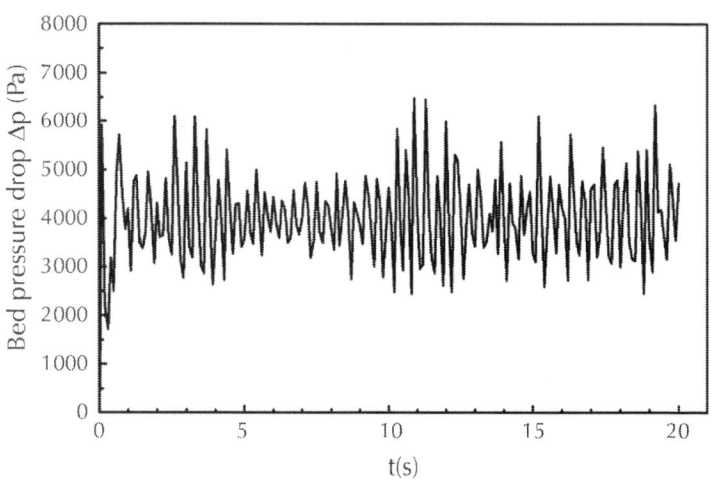

Figure 7: Bed pressure drop Δp against time t. T_r=820 °C, S/B=1.2.

As shown in Fig. 4 and Fig. 5, the vigorous fluidization is characterized by the formation of large bubbles and slugs whose intensive eruptions can make light particles have high velocities and then reach the top outlet where they are eventually entrained out of the reactor (substantiated by snapshots at different times in Fig. 5). Fig. 8 shows the moving trajectory for a selected biomass particle before it is entrained. It is seen that, before entrainment occurs, the particle changes its moving direction and falls back (preferably near the wall) into the bed many times due to gas–particle interactions, particle–particle collisions and boundary effects near the bed top. This mechanism makes biomass particles have a long residence time in the reactor and a high carbon conversion ratio, which favors the syngas production from char gasification.

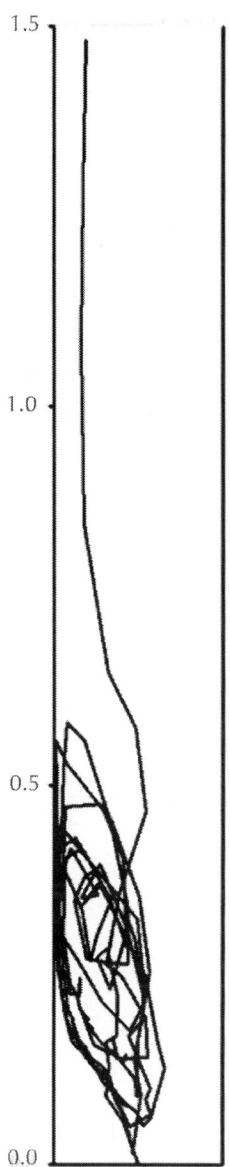

Figure 8: Moving trajectory for a selected biomass particle before it is entrained out of the reactor. T_r=820 °C, S/B=1.2.

Product Gas Composition

For biomass gasification, H_2 and CO are the two most important product gas species. Fig. 9 and Fig. 10 illustrate the H_2 and CO mass fraction distributions in the reactor under base conditions (T_r=820 °C, S/B=1.2), respectively. It can be observed that, at the lower part of the reactor, the concentrations of H_2 and CO are high at similar locations representing regions where the biomass temperature has increased enough to produce large quantities of gas products due to devolatilization and gasification reactions. Moreover, the conditions in the reactor are not symmetrical which is also caused by the gas products from biomass by pyrolysis and gasification. From the analysis in the previous section, we know that, in a vigorous fluidized bed reactor, particles tend to migrate outwards toward the wall, driven by gas–particle interactions, particle–particle collisions and boundary effects, and then descend along the wall. As a result, there is a higher concentration of particles in the wall region where H_2 and CO concentrations are augmented as shown in Fig. 9 and Fig. 10. At the upper part of the reactor, the almost homogeneities in the mass fractions of H_2 and CO are a result of both the lower particle concentration and the gas transport process in the reactor.

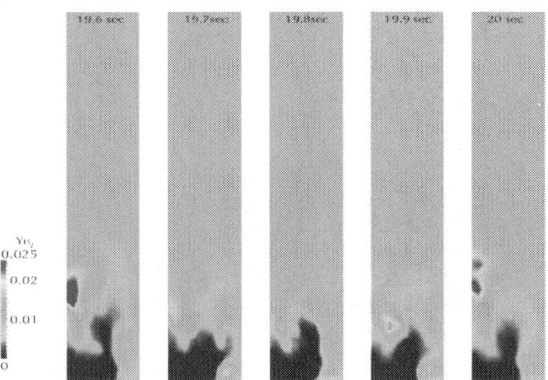

Figure 9: Snapshots of H_2 mass fractions with the time increment being 0.1 s at the end of simulation. T_r=820 °C, S/B=1.2.

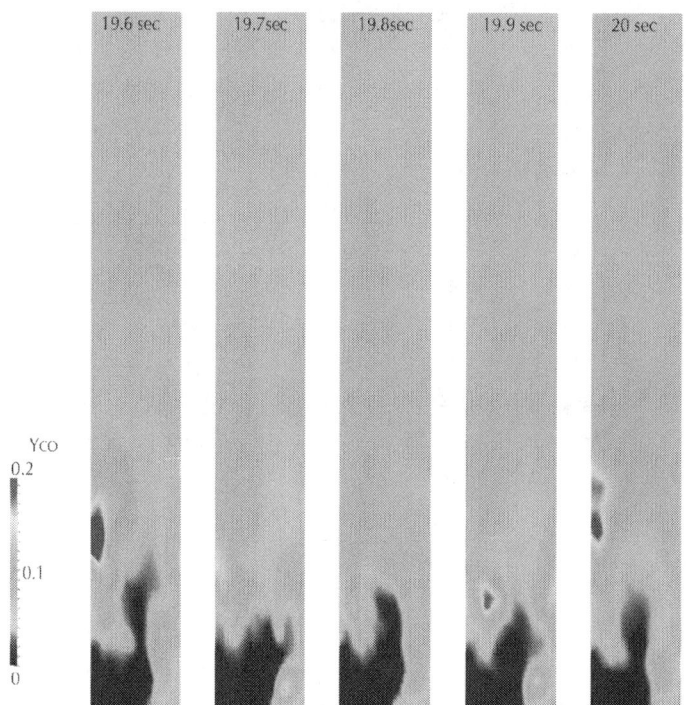

Figure 10: Snapshots of CO mass fractions with the time increment being 0.1 s at the end of simulation. T_r=820 °C, S/B=1.2.

Fig. 11 shows the volume fractions of the product gas compositions at the reactor outlet as a function of time t for the base case (T_r=820 °C, S/B=1.2). Note that the calculated results are based on the dry and N_2 free gas, which is consistent with the experimental study of Song et al. (2012). It is observed that there is only a strong dependence of product gas compositions on t in the initial period of simulation (t<5 s). After the initial period (t>5 s), each composition reaches a quasi-steady state. Thus in the following sections, all the quantitative results are on a time-average basis from t=5 s to 20 s.

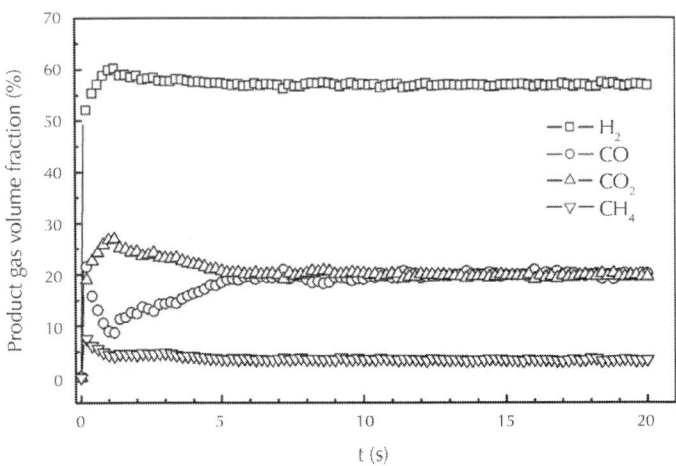

Figure 11: Temporal evolution of product gas volume fractions at the reactor outlet. T_r=820 °C, S/B=1.2.

Effect of Reactor Temperature

Operating rector temperature (T_r) plays an important role in biomass gasification. Fig. 12 shows comparisons of the calculated results with the experimental data of Song et al. (2012) for product gas composition versus reactor temperature in the range of 820–920 °C. The steam/biomass mass ratio (S/B) is fixed at 1.2. It can be observed that, the predictions of the model show good conformance to the experimental measurements. For the two most important syngas species (H_2, CO), the minimum relative error of calculation to experiment is about 1% and the maximum relative error is less than 25%. For CO_2, the maximum relative error is also within 30%. The underestimation of CH_4 can be attributed to the simplification of pyrolysis model and the neglect of tar and methanation reaction. Considering there exist no complete and unified set of gasifier chemistry equations and reaction rates in the open literature, errors cannot be avoided. This implies that the present CFD–DEM simulations are reasonable and the validity of the integrated model is verified.

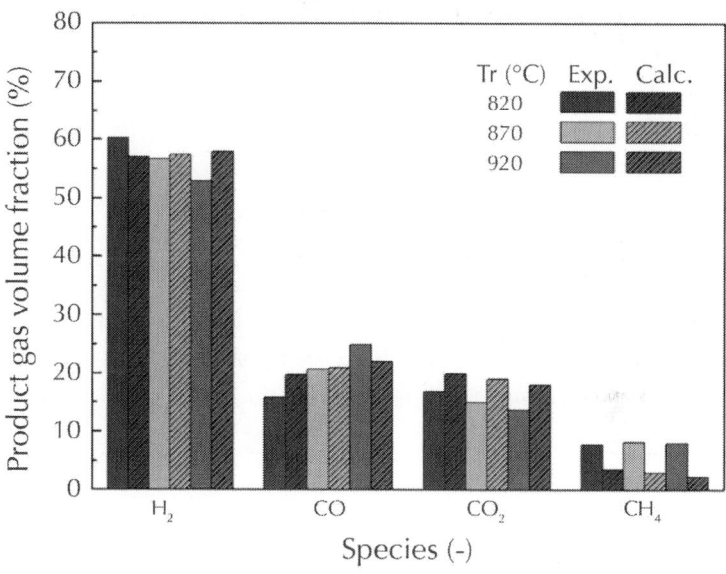

Figure 12: Effect of reactor temperature on product gas composition at the reactor outlet. S/B=1.2.

The product gas composition is the result of the combination of a series of complex and competing reactions, as given in reactions (R1–R4). Generally speaking, higher temperature favors the products in endothermic reactions. Those endothermic reactions include the Boudouard (R1), the (R2) and the methane-steam reforming reaction (R3). Thus reactions (R1)–(R3) are strengthened with an increase in the reactor temperature, which result in an increase of CO and a decrease of CO_2 and CH_4 in the product gas. For H_2, on the one hand, high temperature is in favor of H_2 formation owing to endothermic reactions (R2) and (R3). On the other hand, the temperature increase impels the exothermic water–gas shift reaction (R4) toward the negative direction at the expense of H_2. Therefore, the trend of H_2 content with increasing temperature is governed by the competing reactions (R2)–(R4). As shown in Fig. 12, H_2 content slightly decreases with an increase in the reactor temperature for the experiment, while it is not very sensitive to the temperature change for the simulation.

Effect of Steam/Biomass Mass Ratio

The effect of steam/biomass mass ratio (S/B) on the product gas composition at the reactor temperature of 820 °C is shown in Fig. 13. Again, the calculated exit gas compositions are in a good agreement with the experiment. With the increase of S/B, H_2 and CO_2 concentrations increase while CO concentration decreases. This can be mainly explained by water–gas shift reaction (R4) and high S/B boosts the forward reaction of (R4). Furthermore, due to methane-steam reforming reaction (R3), slightly decreasing trend of CH_4 composition with S/B is observed.

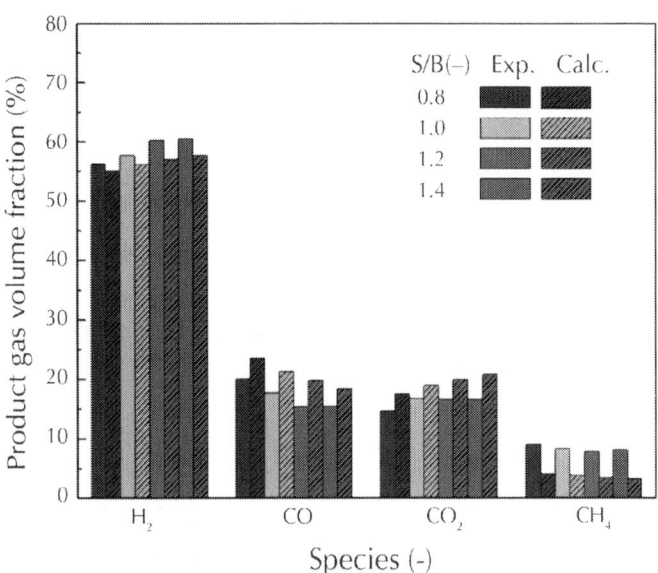

Figure 13: Effect of steam/biomass mass ratio on product gas composition at the reactor outlet. T_r=820 °C.

Effect of Biomass Injection Position

Biomass injection position is another important parameter for design purposes. Fig. 14 shows the effect of three different injection points on the biomass particle distributions. For clarity purpose, the sand particles are excluded in the figure. As shown in Fig. 14, besides the default bottom feed point (Feed1), two other feed points, Feed2 and Feed3, are created at the left side wall and located at 0.2 m and 0.6 m above the bottom of the reactor, respectively. Feed2 denotes a point at the lower part of the sand bed and Feed3 represents a point just above or near the top of the sand bed. Therefore, the three feeding points adopted covers both bottom and top feeding of fuel which are commonly used in practical applications.

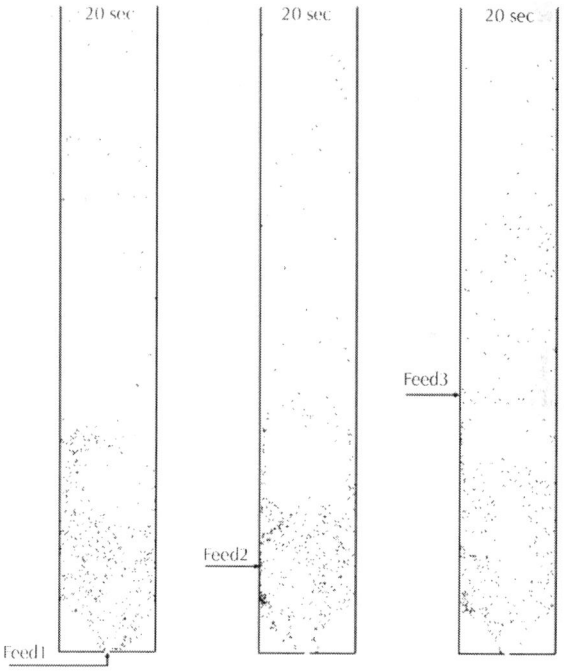

Figure 14: Biomass particle distributions at the end of simulation for three different injection positions. Note that sand particles are excluded for clarity purpose. T_r=820 °C, S/B=1.2.

Fig. 14 shows that, for Feed1 and Feed2, no significant difference related to biomass particle distributions is observed except for a small local accumulation of biomass close to Feed2 position. However, for Feed3 where biomass is injected near the sand bed surface, the relatively low density of biomass precludes its good mixing with the sand bed and more biomass particles tend to be in the freeboard and then have a higher probability of being entrained out of the reactor.

Fig. 15 depicts the average biomass particle temperature for the three different injection points. Specifically, the values of particle temperature for the Feed1, Feed2, and Feed3 are 692.3 °C, 686.9 °C, and 661.3 °C, respectively. As expected, Feed1 has the highest biomass particle temperature due to its best mixing performance.

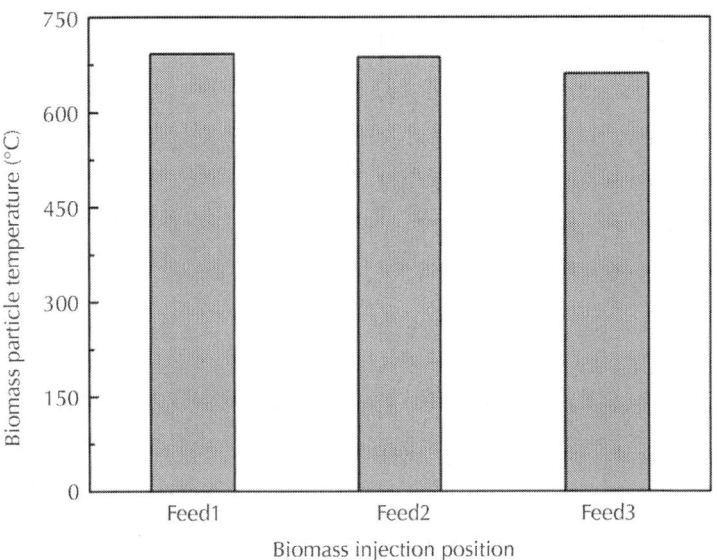

Figure 15: Average biomass particle temperature for the three different injection positions. T_r=820 °C, S/B=1.2.

Fig. 16 shows the average moisture content of biomass particles for the three different injection points. It can be seen that the moisture content is very low for all three injection positions because the vaporization process occurs at a very fast rate due to the high operating temperature (T_r=820 °C). Specifically, the values of moisture content for the Feed1, Feed2, and Feed3 are 0.07%, 0.10%, and 0.20%, respectively. Again, as expected, Feed3 has the highest moisture content due to its worst mixing performance which in turn results in a lowest biomass particle temperature as shown in Fig. 15.

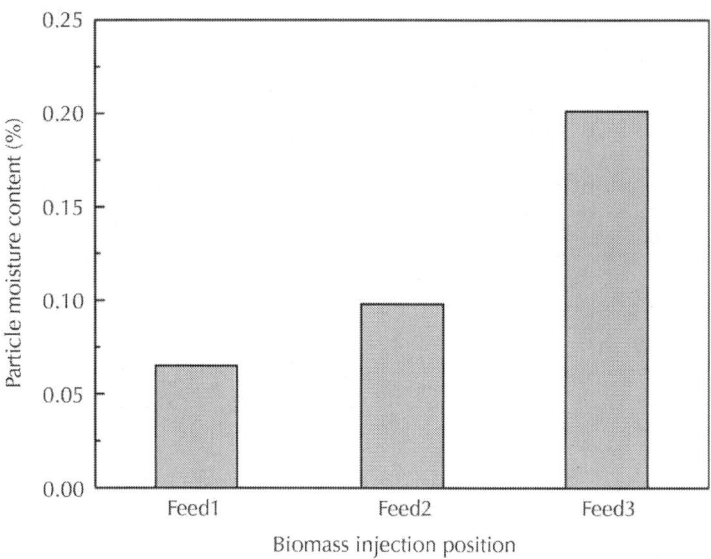

Figure 16: Average moisture content of biomass particles for the three different injection positions. T_r=820 °C, S/B=1.2.

Carbon conversion (CC) is a vital index used for evaluating the performance of gasification. It is defined as follows (Chen et al., 2013),

$$CC \ (\%) = \frac{\dot{m}_{out,CO}\,12/28 + \dot{m}_{out,CO_2}\,12/44 + \dot{m}_{out,CH_4}\,12/16}{\dot{m}_{in,fuel}\,Y_c} \times 100 \tag{23}$$

where Y_c is the mass fraction of carbon in the feed fuel (biomass).

Fig. 17 shows the CC at the reactor outlet for the three injection points. Specifically, the values of CC for the Feed1, Feed2, and Feed3 are 95.3%, 94.9%, and 86.7%, respectively. The CC decreases as the height of injection point increases owing to both an increase of solid entrainment and a decrease of particle residence time and particle temperature (see Fig. 14 and Fig. 15).

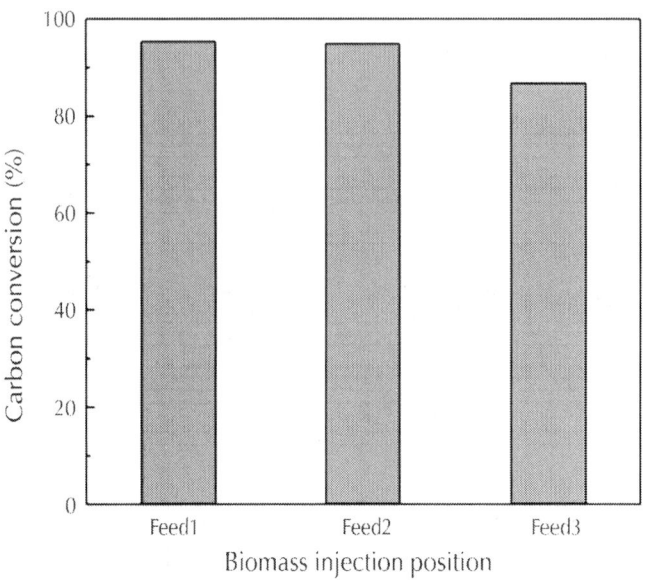

Figure 17: Carbon conversion at the reactor outlet for the three different injection positions. T_r=820 °C, S/B=1.2.

CONCLUSIONS

A comprehensive CFD–DEM numerical model has been developed to simulate the biomass gasification process in a fluidized bed reactor. The gasifying agent is steam. The methodology is based on an Eulerian–Lagrangian concept, which uses an Eulerian method for gas phase and a discrete element method for particle phase. Each particle is individually tracked and associated with a range of physical and thermo-chemical properties, making it possible to look at accurate and detailed multi-scale information (i.e., any desired particle property, trajectory, and particle interaction) over the entire particle life time. The integrated model further considers particle collisions, hydrodynamics of dense gas–particle flow in fluidized beds, turbulence, heat and mass transfer, radiation, particle shrinkage, pyrolysis, as well as homogeneous and heterogeneous chemical reactions. The interaction between the continuous gas phase and the discrete particle phase is also considered by treating the exchange of mass, momentum and energy between the two systems as source terms in the governing equations.

Effects of different operating conditions, such as reactor temperature, steam/biomass mass ratio, and biomass injection position, on the gasification performance are analyzed. Simulation results are analyzed both qualitatively and quantitatively in terms of particle flow pattern, particle mixing and entrainment, bed pressure drop, product gas composition, and carbon conversion. Results show that higher temperatures are favorable for the products in endothermic reactions (e.g. H_2 and CO). With the increase of steam/biomass mass ratio, H_2 and CO_2 concentrations increase while CO concentration decreases. The carbon conversion decreases as the height of injection point increases owing to both an increase of solid entrainment and a decrease of particle residence time and particle temperature. Meanwhile, the integrated model has also been validated by comparing the calculated results with the experimental data. This indicates that the proposed CFD–DEM model can provide not only the macro structures at fluidized bed scale (bubble or slug) but also detailed microscopic information at the particle level (such as particle trajectory, particle–particle interaction, particle entrainment, and particle reaction, see Fig. 5, Fig. 8 and Fig. 14) which is impossible to obtain by an Eulerian–Eulerian approach. So our proposed model can be a powerful tool to gain an insight into the complex dense gas–particle flow behaviors and chemical reaction characteristics simultaneously in the process of biomass gasification in a fluidized bed reactor.

ACKNOWLEDGMENTS

The authors would like to thank partners in CenBio, the BioEnergy Innovation Centre, and GasBio for financial support.

REFERENCES

1. Abani,N.,Ghoniem,A.F.,2013.Largeeddysimulationsofcoalgasification inan entrained flow gasifier. Fuel104,664–680.
2. Backreedy,R.I.,Fletcher,L.M.,Ma,L.,Pourkashanian,M.,Williams ,A.,2006. Modelling pulverisedcoalcombustionusingadetailed-coalcombustionmodel. Combust. Sci.Technol.178,763–787.
3. Boyalakuntla,D.S.,2003.SimulationofGranularandGas–Solid FlowsusingDiscrete Element Method (Ph.D.thesis).CarnegieMell onUniversity,Pittsburgh, Pennsylvania.
4. Bruchmüller, J.,vanWachem,B.G.M.,Gu,S.,Luo,K.H.,Brown,R .C.,2012.Modeling the thermochemicaldegradationofbiomassin-sideafastpyrolysisfluidizedbedreactor.AIChEJ.58(10),3030–3042.
5. Chen, W.,Chen,C.,Hung,C.,Shen,C.,Hsu,H.,2013.Acompari-sonofgasification phenomena amongrawbiomass,torrefied bio-massandcoalinanentrained- flow reactor.Appl.Energy112, 421–430.
6. Cundall, P.A.,Strack,O.D.L.,1979.Adiscretenumericalmodelfor-granular assemblies. Geotechnique29,47–65.
7. Ergun, S.,1952.Fluid flow throughpackedcolumns.Chem.Eng. Prog.48,89–94.
8. Ergüdenler, A.,Ghaly,A.E.,Hamdullahpur,F.,Al-Taweel,A.M.,1997. Mathematical modeling ofa fluidized bedstrawgasifier: PartI—model development.Energy Source 19,1065–1084.
9. Gerber, S.,Behrendt,F.,Oevermann,M.,2010.AnEulerianmodelin-gapproachof wood gasification inabubbling fluidized bedreacto-rusingcharasbed material. Fuel89,2903–2917.
10. Gerber, S.,Oevermann,M.,2014.AtwodimensionalEuler–Lagrang-ianmodelof wood gasification inacharcoalbed – Part I:model de-scriptionandbase scenario. Fuel115,385–400.
11. Gidaspow, D.,1994.Multiphase Flow and Fluidization.Academic Press,SanDiego, USA.
12. Gil, J.,Corella,J.,Aznar,M.P.,Caballero,M.A.,1999.Biomassgasifi-cation inatmo- spheric andbubbling fluidized bed:effectofthety peofgasifyingagentonthe product distribution.BiomassBioener-gy17,389–403.
13. Gómez-Barea, A,Leckner,B.,2010.Modelingofbiomassgasifica-tion in fluidized bed. Prog.EnergyCombust.36,444–509.
14. Hoomans, B.P.B.,Kuipers, J.A.M, Briels, W.J., van Swaaij, W.P.M.,

1996. Discrete particle simulationofbubbleandslugformationinatwo-dimensionalgas fluidised bed:ahard-sphereapproach.Chem. Eng.Sci.51,99–118.

15. Jones, W.P.,Lindstedt,R.P.,1988.Globalreactionschemesforhydrocarboncombus- tion. Combust.Flame73,233–249.

16. Kafui, K.D.,Thornton,C.,Adams,M.J.,2002.Discreteparticle-continuum fluid modelling ofgas–solid fluidised beds.Chem.Eng. Sci.57,2395–2410.

17. Kern,S.,Pfeifer,C.,Hofbauer,H.,2013.Gasification ofwoodinadual fluidized bed gasifier: influence offuelfeedingonprocessperformance.Chem.Eng.Sci.90, 284–298.

18. Kim, Y.D.,Yang,C.W.,Kim,B.J.,Kim,K.S.,Lee,J.W.,Moon,J.H.,Yang ,W.,Yu,T.U., Lee, U.D,2013.Air-blowngasification ofwoodybiomassinabubbling fluidized bed gasifier. Appl.Energy112,414–420.

19. Ku,X.,Li,T.,Løvås,T.,2013.Influence ofdragforcecorrelationsonperiodic fluidization behaviorinEulerian–Lagrangian simulationofabubbling fluidized bed. Chem.Eng.Sci.95,94–106.

20. Kumar,M.,Ghoniem,A.F.,2012.Multiphysicssimulationsofentrained flow gasification. PartII:constructingandvalidatingtheover allmodel.EnergyFuels 26, 464–479.

21. Lathouwers, D.,Bellan,J.,2001.Modelingofdensegas-solidreactive mixtures applied tobiomasspyrolysisina fluidized bed. Int.J.Multiph.Flow27(12), 2155–2187.

22. Li, X.T.,Grace,J.R.,Lim,C.J.,Watkinson,A.P.,Chen,H.P.,Kim,J.R., 2004. Biomass gasification inacirculating fluidized bed.Biomass-Bioenergy26,171–193.

23. Liu, D.,Chen,X.,Zhou,W.,Zhao,C.,2011.Simulationofcharandpropane combustion ina fluidized bedbyextendingDEM–CFD approach.Proc.Combust. Inst. 33,2701–2708.

24. Meng, X.,deJong,W.,Fu,N.,Verkooijen,A.H.M.,2011.Biomassgasification ina 100kWthsteam-oxygenblowncirculating fluidized bedgasifier: effectsof operationalconditionsonproductgasdistributionandtarformation.Biomass Bioenergy 35,2910–2924.

25. Nikoo, M.B.,Mahinpey, N.,2008. Simulation of biomass gasification in fluidized bed reactorusingASPENPLUS.BiomassBioenergy32(12),1245–1254.

26. OpenCFD Ltd,2012.OpenFOAM-TheopensourceCFDtoolboxuserguide(Version 2.1.1) http://www.openfoam.org/docs/ .

27. Papadikis, K.,Gu,S.,Bridgwater,A.V.,2010.ACFDapproachonthe

effect of particle size oncharentrainmentinbubbling fluidised bed-reactors.BiomassBioenergy 34, 21–29.

28. Prakash,N.,Karunanithi,T.,2008.Kineticmodelinginbiomasspyrol-ysis – a review. J. Appl.Sci.Res.4(12),1627–1636.

29. Qin, K.,Jensen,P.A.,Lin,W.,Jensen,A.D.,2012.Biomassgasification behaviorinan entrained flow reactor:gasproductdistributionandso otformation.Energy Fuels 26,5992–6002.

30. Sadaka, S.S.,Ghaly,A.E.,Sabbah,M.A.,2002.Twophasebiomassair-steam gasification modelfor fluidized bedreactors:PartI—model development. Biomass Bioenergy22,439–462.

31. Shen,   L.,Gao,Y.,Xiao,J.,2008.Simulationofhydrogenproduction-frombiomass gasification ininterconnected fluidized beds.Bio-massBioenergy32,120–127.

32. Snider,   D.M.,Clark,S.M.,O'Rourke,P.J.,2011.Eulerian–Lagrangian methodforthree- dimensional thermalreacting flow withapplica-tiontocoalgasifiers. Chem. Eng. Sci.66,1285–1295.

33. Song,T.,Wu,J.,Shen,L.,Xiao,J.,2012.Experimentalinvestigationon-hydrogen productionfrombiomassgasification ininterconnected fluidized beds.Biomass Bioenergy 36,258–267.

34. Taghipour,F.,Ellis,N.,Wong,C.,2005.Experimentalandcomputa-tionalstudyof gas–solid fluidized bedhydrodynamics.Chem.Eng. Sci.60,6857–6867.

35. Tsuji,Y.,Kawaguchi,T.,Tanaka,T.,1993.Discreteparticlesimulation-oftwo-dimensional fluidized bed.PowderTechnol.77,79–87.

36. Tsuji,Y.,Tanaka,T.,Ishida,T.,1992.Lagrangiannumericalsimulation-ofplug flow of cohesionless particlesinahorizontalpipe.Powder-Technol.71,239–250.

37. Wang,X.,Jin,B,Zhong,W.,2009.Three-dimensionalsimulationof fluidized bed coal gasification. Chem.Eng.Process.48,695–705.

38. Warnecke,R.,2000.Gasification ofbiomass:comparisonof fixed-bedand fluidized bed gasifier. BiomassBioenergy18,489–497.

39. Wen,C.Y.,Yu,Y.H.,1966.Mechanicsof fluidization. Chem.Eng. Prog.Symp.Ser.62, 100–111.

40. Xie, J.,Zhong,W.,Jin,B.,Shao,Y.,Huang,Y.,2013.Eulerian–Lagrang-ian methodfor three-dimensional simulationof fluidized bedcoal-gasification. Adv.Powder Technol.24,382–392.

41. Xu,B.H.,Yu,A.B.,1997.Numericalsimulationofthegas–solid flow ina fluidized bed bycombiningdiscreteparticlemethodwithcom-putational fluid dynamics. Chem. Eng.Sci.52,2785–2809.

Citations

CHAPTER 1

Yongzhuo Liu, Weihua Jia, Qingjie Guo, Hojung Ryu, Effect of Gasifying Medium on the Coal Chemical Looping Gasification with CaSO4 as Oxygen Carrier, Chinese Journal of Chemical Engineering, Volume 22, Issues 11–12, November 2014, Pages 1208-1214, ISSN 1004-9541, http://dx.doi.org/10.1016/j.cjche.2014.09.011.

CHAPTER 2

Jakkapong Udomsirichakorn, Prabir Basu, P. Abdul Salam, Bishnu

Acharya, CaO-based chemical looping gasification of biomass for hydrogen-enriched gas production with in situ CO2 capture and tar reduction, Fuel Processing Technology, Volume 127, November 2014, Pages 7-12, ISSN 0378-3820, http://dx.doi.org/10.1016/j.fuproc.2014.06.007.

CHAPTER 3

Seán T. Mac an Bhaird, Phil Hemmingway, Eilín Walsh, Amado L. Maglinao, Sergio C. Capareda, Kevin P. McDonnell, Bubbling fluidised bed gasification of wheat straw–gasifier performance using mullite as bed material, Chemical Engineering Research and Design, Volume 97, May 2015, Pages 36-44, ISSN 0263-8762, http://dx.doi.org/10.1016/j.cherd.2015.03.010.

CHAPTER 4

Calin-Cristian Cormos, Ana-Maria Cormos, Letitia Petrescu, Assessment of chemical looping-based conceptual designs for high efficient hydrogen and power co-generation applied to gasification processes, Chemical Engineering Research and Design, Volume 92, Issue 4, April 2014, Pages 741-751, ISSN 0263-8762, http://dx.doi.org/10.1016/j.cherd.2013.08.023.

CHAPTER 5

Ulises Badillo-Hernandez, Luis Alvarez-Icaza, Jesus Alvarez, Model design of a class of moving-bed tubular gasification reactors, Chemical Engineering Science, Volume 101, 20 September 2013, Pages 674-685, ISSN 0009-2509, http://dx.doi.org/10.1016/j.ces.2013.07.001.

CHAPTER 6

Arturo Gomez, Nader Mahinpey, A new method to calculate kinetic parameters independent of the kinetic model: Insights on CO2 and steam gasification, Chemical Engineering Research and Design, Volume 95, March 2015, Pages 346-357, ISSN 0263-8762, http://dx.doi.org/10.1016/j.cherd.2014.11.012.

CHAPTER 7

Xiaoke Ku, Tian Li, Terese Løvås, CFD–DEM simulation of biomass gasification with steam in a fluidized bed reactor, Chemical Engineering Science, Volume 122, 27 January 2015, Pages 270-283, ISSN 0009-2509, http://dx.doi.org/10.1016/j.ces.2014.08.045.

Index

A

Acid Gas Removal (AGR) 76, 83
Activation energy 151, 154, 155, 156,
    157, 158, 159, 162, 164, 165, 170,
    171, 174, 177, 178, 179, 183, 184
Air Separation Unit (ASU) 81

C

Carbon capture 70, 71, 72, 75, 76, 81, 89,
    92, 93, 94, 95, 97, 98, 99, 101, 102
Carbon capture and storage (CCS) 69
Catalytic steam 27
Chemical looping combustion (CLC) 3
Chemical looping gasification (CLG) 3,
    24

Chemical reaction 74
Circulating fluidized bed (CFB) 26
Coal gasification 2, 4
Cold gas efficiency (CGE) 51, 59
Computational fluid dynamic (CFD)
    191
Computational fluid dynamics (CFD)
    105

D

Derivative thermogravimetric analysis
    (DTG) 12
Discrete element method (DEM) 189,
    191

E

Energy Policy & Planning Office (EPPO) 40
Enhanced Oil Recovery (EOR) 80
Equivalence ratio (ER) 47

F

Finite-difference (FD) 106

G

Gasification of biomass 24, 27, 36, 38, 41, 42, 43, 228
Gas yield 24, 33, 36, 38
Global warming 24

H

Higher heating value (HHV) 28
Hot and cold composite curves (HCC and CCC) 87

I

Integrated core model (ICM) 158

L

Langmuir–Hinshelwood (LH) 155
Lower heating value (LHV) 51, 80

N

N continuously stirred tank reactors (CSTRs) 106

O

ordinary differential equations (ODEs), 106

P

Partial oxidation 152, 153

Q

Quasi-steady state (QSS) 104, 106

R

Random pore model (RPM) 153

S

Soave–Redlich–Kwong (SRK) 84
Sorption enhanced water gas shift – SEWGS 73
Standard conditions of temperature and pressure (STP) 62
Standard temperature and pressure (STP) 46

T

Thermal gravimetric analyzer (TGA) 7
Thermochemical conversion technologie 153
Tri-ethylene-glycol (TEG) 84